Elementary Math with Pizzazz!

Grade 3

Wright Group
McGraw-Hill

Acknowledgments

Project Editors ► Judith Mendoza, Diane Nieker

Writers ► Barbara Bobb, Judith Mendoza,
Diane Nieker, Warren Traver

Design Manager ► Amy O'Brien Krupp

Design ► Gerta Sorensen

Interior Illustrations ► Lindaanne Donohoe

Composition ► Sorensen London, Inc.

One Prudential Plaza
Chicago, IL 60601

ISBN: 0-3220-9690-1
Customer Service 800-624-0822
www.creativepublications.com

1 2 3 4 5 6 7 8 MAL 08 07 06 05 04 03

Table of Objectives

Numeration and Place Value

Operations with Numbers

Measurement

Geometry

Algebra

Data and Probability

About *Elementary Math with Pizzazz!*™

Students need ongoing practice to reinforce learning and to keep skills fresh. *Elementary Math with Pizzazz!* is a series for Grades 1–5 that provides important math concepts in puzzle-like formats to make practice fun.

This series is designed to complement your school's math program. The skills presented at each level were specifically selected to align with the objectives included by many states in their State Standards. All five content areas identified by the National Council for Teachers of Mathematics (NCTM) are addressed at each grade. The activity pages are organized by content area.

You will find *Elementary Math with Pizzazz!* to be a valuable companion to your math program.

- **The activity pages are correlated to grade-level objectives.**

A list of objectives, along with the specific puzzle pages that deal with that objective, can be found on pages iii–iv. In addition, the objective addressed by each page is clearly indicated on the page.

Every puzzle page addresses a skill that students at the grade level should be able to complete. The exercises are intended for students who have already had experience with the particular concept being addressed; some pages present basic skills practice while others provide a greater challenge.

- **The activity pages can be used in a variety of ways.**

Use *Elementary Math with Pizzazz!* to follow up classroom instruction of a new skill, as homework, or to offer students continuing practice and reinforcement at any time throughout the school year. Based on the needs of your students, pages can be assigned to your whole class, to small groups, or as an activity to be completed by individual students. The blackline master format of this series makes it easy to produce the exact number of copies required.

- **The activity pages are easy to use.**
  The directions for each page are clear and easy to understand, making it possible for students to work with minimal teacher instruction. Should you want to work through the first problem with your students to get them started, or give more attention to a particular set of problems, an overhead transparency created from the blackline master can be a valuable visual aid.

- **The activity pages are self-correcting.**
  In most instances, students will know immediately if they have worked problems correctly because their answers will produce a solution for the puzzle. If errors have been made, the puzzle will not be correctly solved and students will be able to identify and rework any problems they missed. The self-correcting design of the pages will result in far less time spent grading papers. In the few instances where pages are not self-correcting, answers are easy to check—usually by visual clues. For your convenience, you'll find a complete answer key beginning on page 73.

Each page in *Elementary Math with Pizzazz!* complements a mathematics objective found in your mathematics curriculum. Whether used for practice, reinforcement, or enrichment, finding answers to riddles and puzzles will motivate students and reward them for completing the exercises. Your students will enjoy a sense of accomplishment as they apply their skills and correctly solve the puzzles. And you will enjoy knowing you are able to offer your students skills practice that is appropriate to their needs.

Name ____________________

Blast Off!

Follow each direction as you advance through the maze.

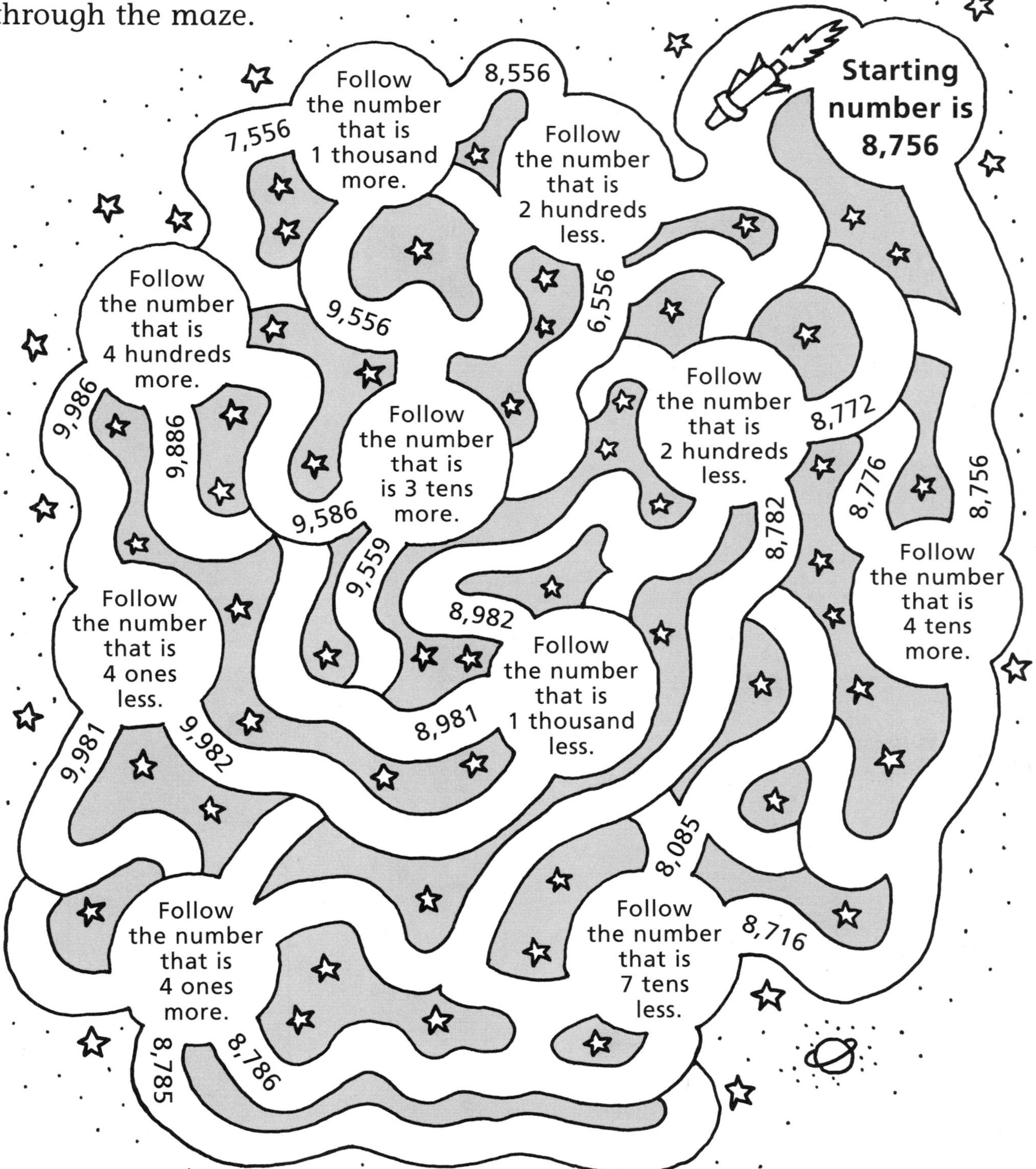

Name ____________________

How Is Your Aim?

Use a ruler to draw a line from the tip of each arrow to the center of the target with the same number. Your lines will cross out letters. Write the letters that are left in order at the bottom of the page.

Ten thousand, one hundred ten →

Eleven thousand, one hundred eleven →

Ten thousand, forty-five →

Twenty-five thousand, four hundred twelve →

One thousand, three hundred →

Seventeen thousand, one hundred seventeen →

Five hundred twenty-two →

One thousand, seven hundred seventeen →

Fifty-six thousand, five hundred twenty-two →

Five thousand, six hundred fifty-two →

Y O U

S A R C

T E D R

I B G D

H T F N

C O N A

R T A R

G E

I T K

Targets: 1,717; 25,412; 522; 10,045; 17,117; 1,300; 11,111; 10,110; 5,652; 56,522

____.

Name ______________________________

The "Write" Numbers

Write a numeral for each number word.
Then color the section of the design that matches your numeral.

1. Seventeen thousand, one ______________
2. Forty-five thousand, seven hundred forty-one ______________
3. Twenty-six thousand, four hundred fifty-three ______________
4. Ninety-five thousand, five ______________
5. Sixty thousand, one hundred three ______________
6. Nine thousand, ten ______________
7. Twenty-one thousand, six hundred fifty-four ______________
8. Nineteen thousand, thirty ______________
9. Six thousand, fifty-two ______________
10. Twenty-five thousand, one hundred three ______________
11. Sixty thousand, five hundred fifty-five ______________
12. Forty thousand, two hundred thirty-four ______________

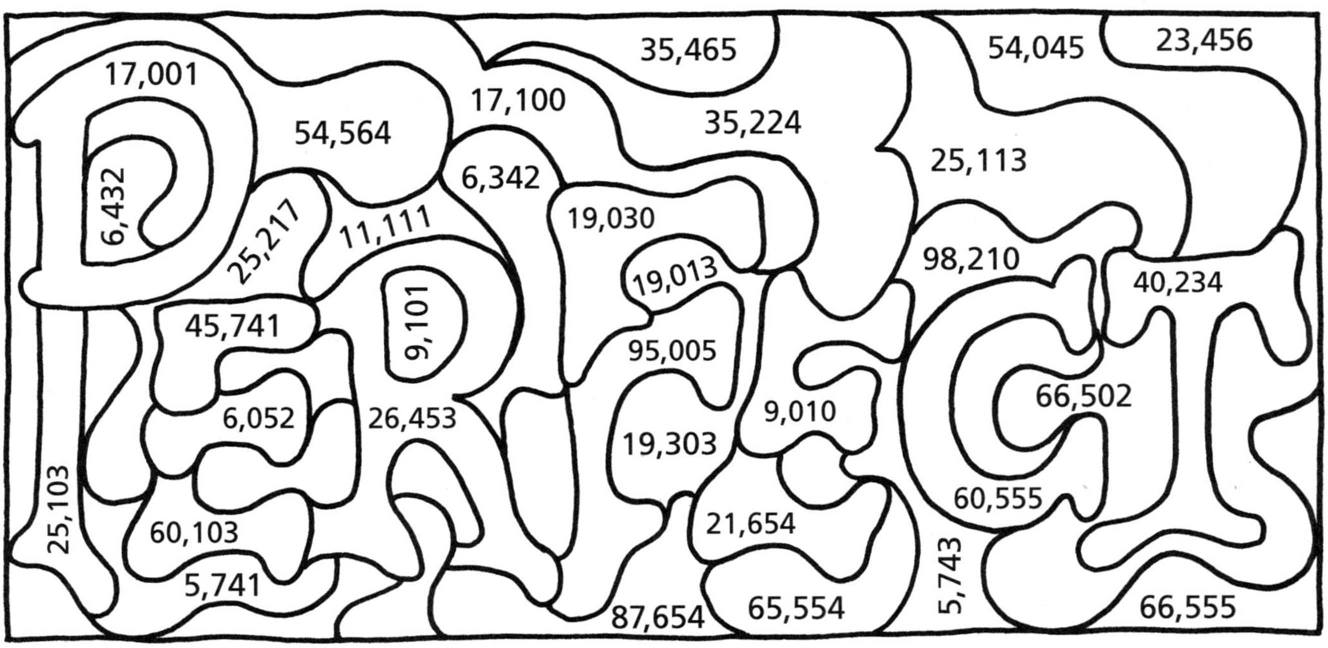

Name ______________________

How Hard Was It to Decode This Message?

Write the number.

176,142 + 5 hundreds

Eight hundred twelve thousand, three hundred

Eighteen thousand, two hundred fifty-four

Nine thousand, three hundred twenty-three

3 hundreds + 8 tens + 3 ones

A	E	I	O	B	T
1	7	6	6	4	2

623,211 + 6 ten thousands

One hundred fifty-one thousand, one hundred

Thirty-five thousand, seven hundred seventeen

Eighty thousand, six hundred seventy-eight

632 + 4 tens

H	C	W	Y	D	S

Find new numbers by reading the columns.
Match the letter of the column to the number in the code.

___ ___ ___ ___   ___ ___ ___   ___ ___
20,433 61 6,289 10,782   3,150 18 10,782   18 10,782

___ ___ ___ ___   ___ ___   ___ - ___ - ___
711 18 10,782 21,766   18 10,782   18 40,528 8,538

___ ___   ___ ___ ___ ___ ___ ___ .
20,433 63,233   10,177 711 8,538 63,233 10,177 711

Name ______________________________

Mole Hole

Follow the directions in each circle to help the mole find his home.

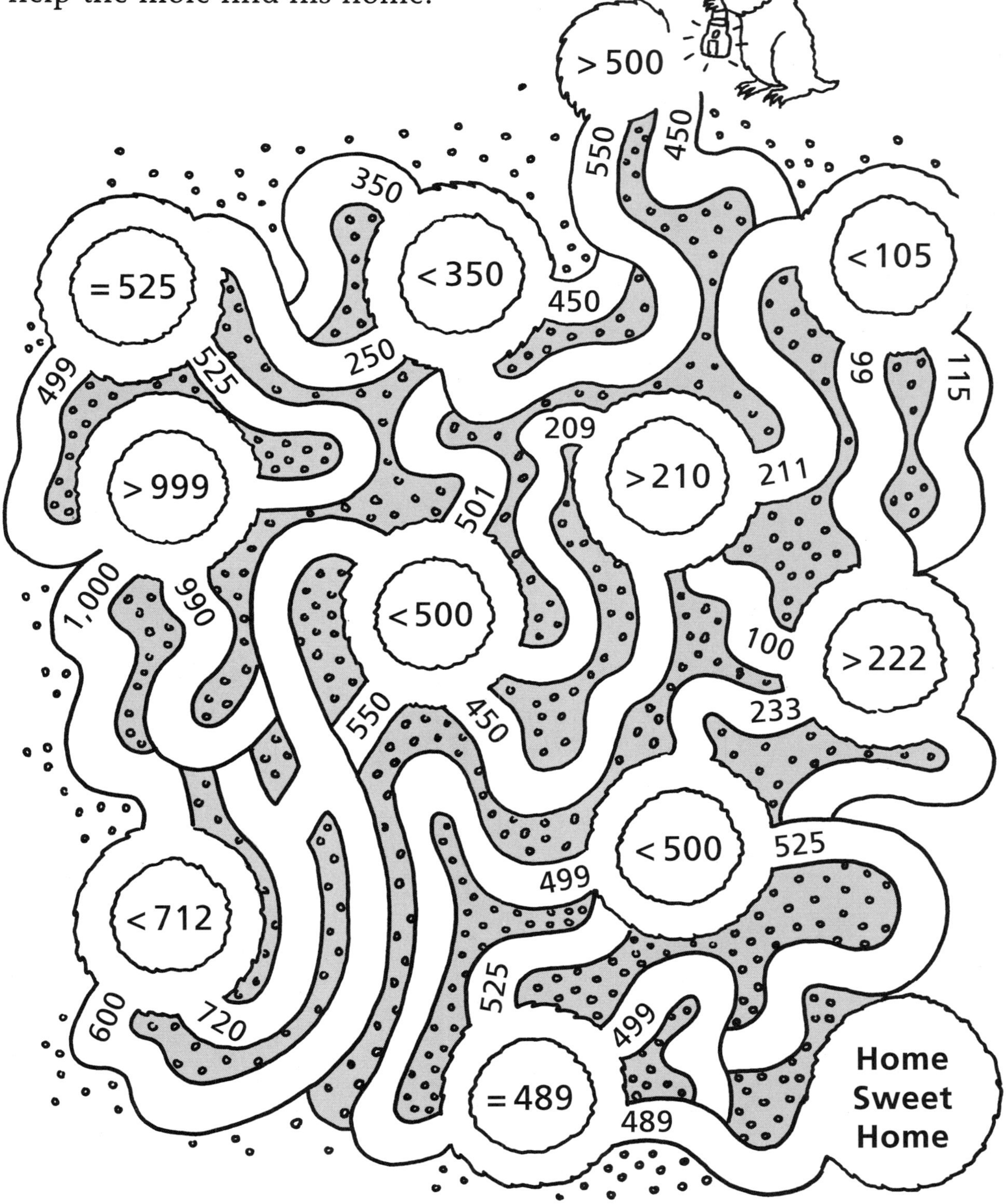

Name ____________________

An Evening Song

Fill in the ◯ to compare each pair of numbers.
Use < for less than, > for greater than, or = for equal.

Color spaces with > black.
Color spaces with < brown.
Color spaces with = yellow.

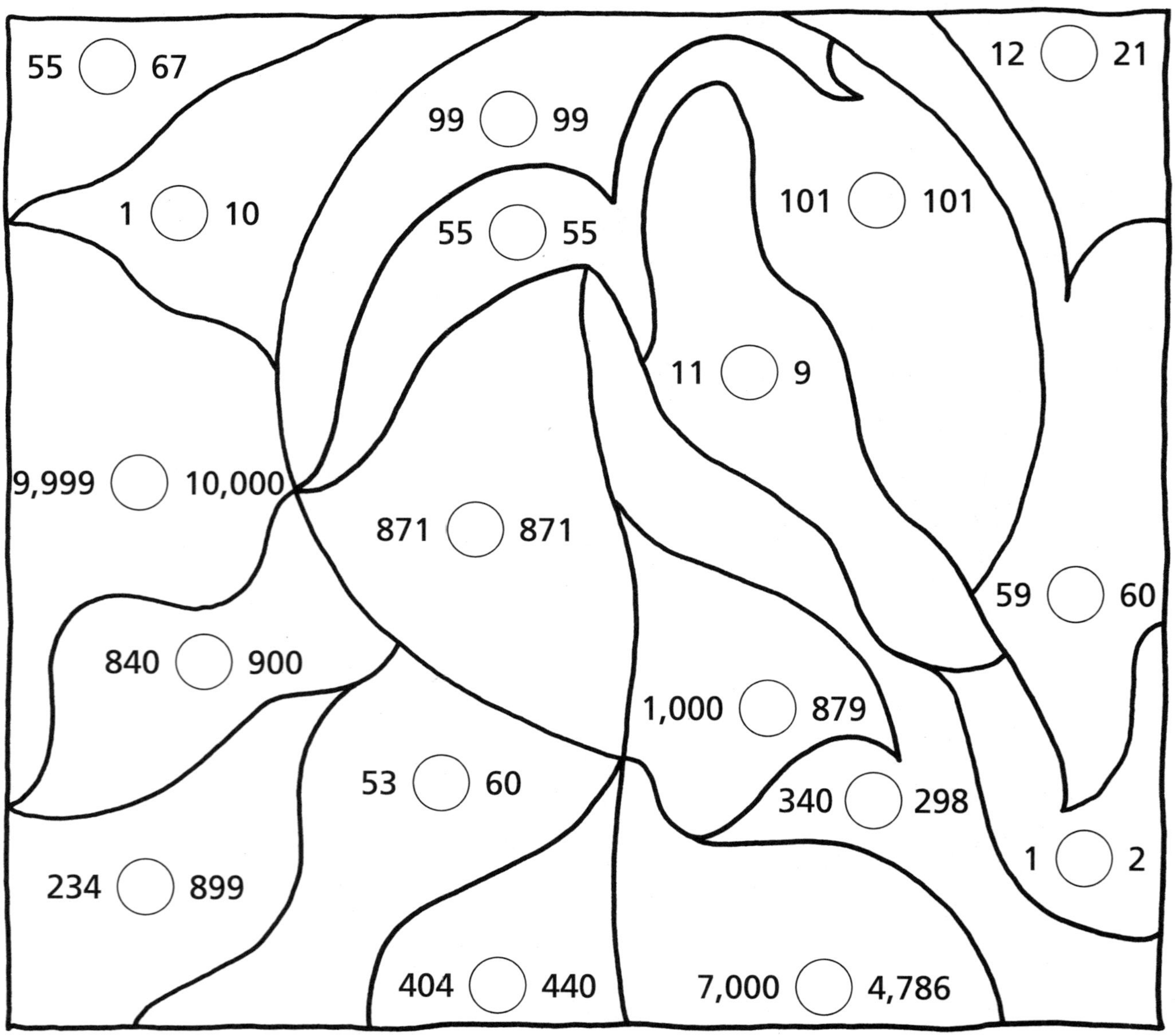

Name ______________________

Animal Abilities

Write the missing whole number. Then find the number in the code and write the letter above it.

1. I'm the number just before 100. ________ **N**
2. You'll find me between 999 and 1,001. ________ **H**
3. When you count, you say me right before 10,000. ________ **W**
4. I'm the 6th counting number you say after 595. ________ **T**
5. I'm the number between 817 and 819. ________ **O**
6. I come right before 889. ________ **E**
7. If you count back 2 numbers from 9,911, you'll find me. ________ **I**
8. I'm the number between 609 and 611. ________ **P**
9. You'll find me right before 910. ________ **G**
10. I'm the 3rd counting number you say after 805. ________ **S**

How does a skunk smell?

___ ___ ___ ___   ___ ___ ___   ___ ___ ___ ___

9,999 9,909 601 1,000   1,000 9,909 808   99 818 808 888

What do pigs use when they write letters?

___ ___ ___   ___ ___ ___ ___

610 9,909 909   610 888 99 808

Name ______________________

Leaping Lily Pads!

Round each number.

1. 156 to the tens place ____________
2. 329 to the nearer hundred ____________
3. 360 to the hundreds place ____________
4. 72 to the nearer ten ____________
5. 277 to the tens place ____________
6. 609 to the hundreds place ____________
7. 609 to the tens place ____________
8. 691 to the tens place ____________
9. 199 to the nearer hundred ____________
10. 8 to the nearer ten ____________
11. 18 to the tens place ____________
12. 61 to the hundreds place ____________

Complete the dot-to-dot. Follow your answers in order.

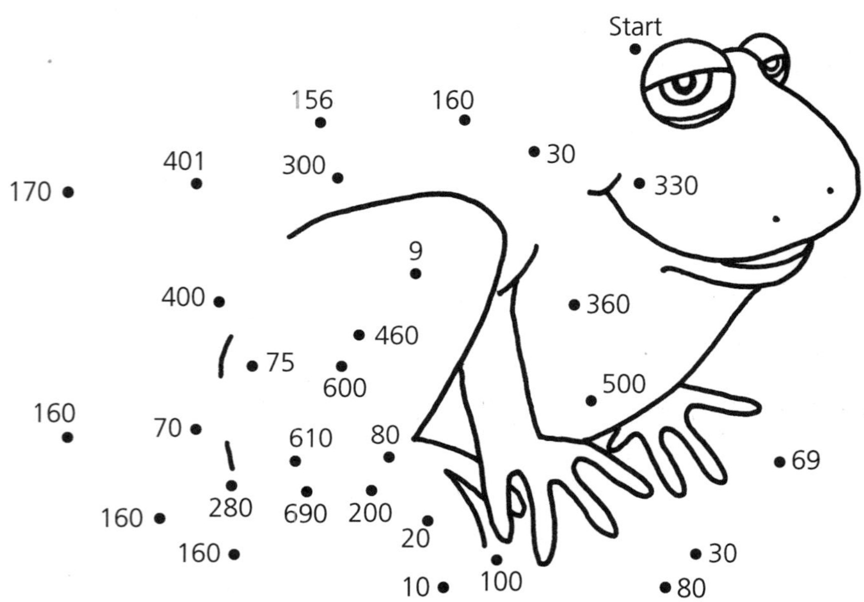

Name ________________________

Estimation Puzzle

Note: Use with page 10.

Write the front-end estimate for each number.

14 → ________ 

72 → ________ 

604 → ________ 

64 → ________ 

26 → ________ 

999 → ________ 

829 → ________ 

32 → ________ 

94 → ________ 

433 → ________ 

362 → ________ 

735 → ________ 

56 → ________ 

156 → ________ 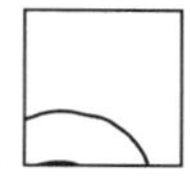

211 → ________ 

85 → ________ 

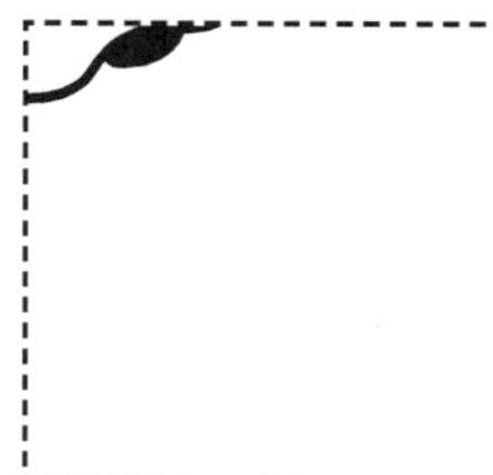

Cut out these 16 squares to use on page 10.

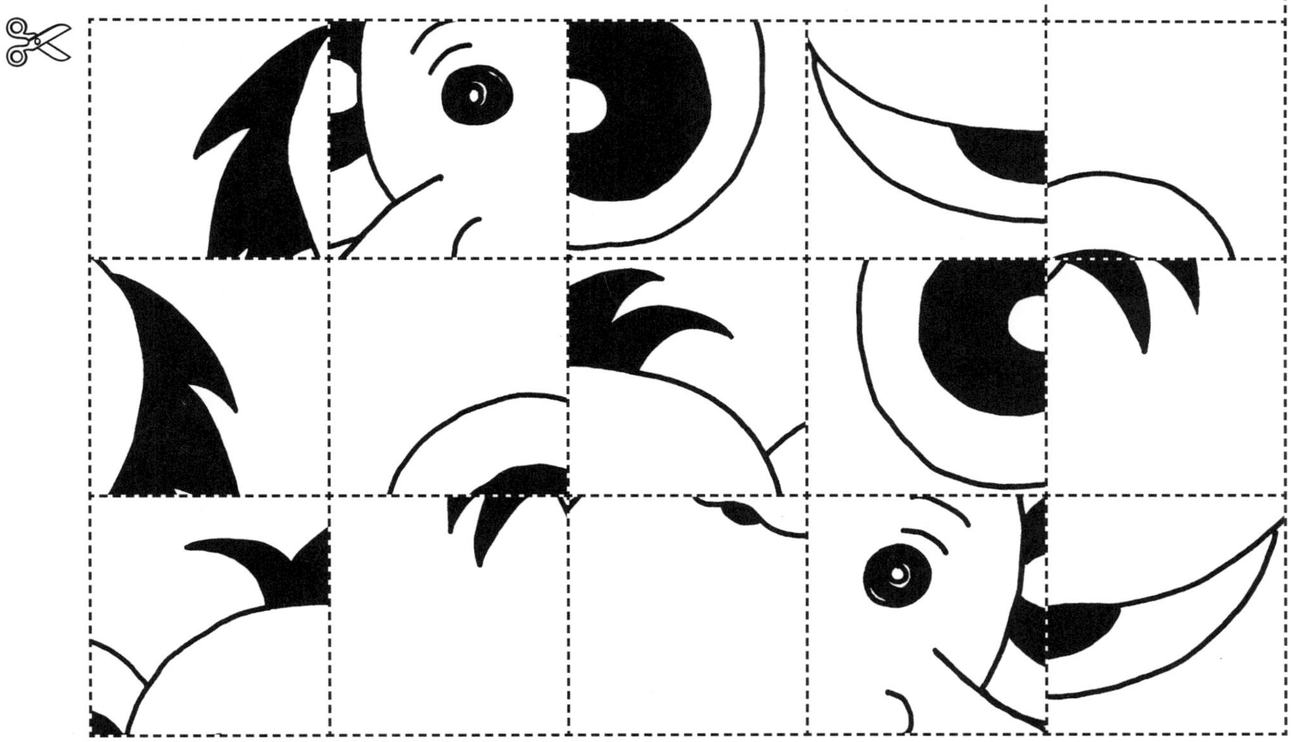

Name ____________________

Estimation Puzzle

Note: Use with page 9.

Match each cutout square to one pictured on page 9.
Find the front-end estimate for that picture in the grid.
Paste the cutout in the square with the estimate.

900	90	50	100
600	800	20	300
700	70	30	80
10	60	400	200

▶ Practice addition and subtraction facts

Name ____________________

Cross-Number Puzzles

Write a number in each box with a letter to make true sentences that go across and down.

14	−	8	=	R
+		−		+
2	+	O	=	4
=		=		=
S	−	6	=	10

3	+	Z	=	10
+		+		+
4	+	V	=	5
=		=		=
7	+	U	=	15

13	+	6	=	I
+		−		+
A	−	2	=	1
=		=		=
16	+	4	=	H

D	−	6	=	12
−		+		−
4	+	8	=	Q
=		=		=
14	−	F	=	0

9	+	9	=	18
+		−		−
Y	+	T	=	N
=		=		=
14	−	5	=	9

2	+	8	=	E
+		−		+
L	−	6	=	7
=		=		=
15	+	2	=	P

Find each answer in the code. Write its letter above it.

5 2 8   16 2 13 1 10 18

4 20 10 16 10   17 8 7 7 13 10 16

14 3 19 6   3 9 18   16 12 8 3 6 10.

Name ___________________________

Dot-to-Dot

Solve each exercise. Then complete the dot-to-dot by connecting your answers in order.

1. 18 ÷ 9 = ______ **2.** 3 × 6 = ______ **3.** 7 × 2 = ______

4. 20 ÷ 2 = ______ **5.** 16 ÷ 2 = ______ **6.** 18 ÷ 3 = ______

7. 4 × 5 = ______ **8.** 4 × 4 = ______ **9.** 3 × 4 = ______

10. 16 ÷ 4 = ______ **11.** 9 ÷ 3 = ______ **12.** 5 × 3 = ______

13. 3 × 3 = ______ **14.** 3 ÷ 3 = ______

Name ______________________________

A Day at the Carnival

Solve each problem. Write the boxed word from the question above its answer in the code.

Ferris Wheel
5 tickets
per ride

1. Luis, Sophia, and Bihn purchased tickets [at] the carnival. Luis bought 6 tickets, Sophia bought 7, and Bihn bought 5. They decided to put their tickets together. How many tickets did the three friends have in all? ______ tickets

2. If [the] three friends shared the tickets equally, how many would each one get? ______ tickets

3. Luis held the tickets for everyone. He had a hole in his pocket and 4 tickets fell out. How many [were] left? ______ tickets

4. [When] they discovered that Luis had lost the 4 tickets, Sophia used the rest of her money to buy 2 more tickets. How many tickets did they have then? ______ tickets

5. The three friends decided to ride the Ferris wheel. How many tickets did [they] have to use? ______ tickets

6. How many tickets were left at the [end] of the day? ______ ticket

When did the three friends find the lines for the rides were the longest?

______	______	______	______	______	______
16	15	14	18	6	1

Name ______________________________

Where Do Fish Keep Their Money?

Complete each number sentence.

E $20 \div 5 \times 2 =$ ______

N $16 \div 4 \times 3 =$ ______

V $4 \times 3 + 4 =$ ______

K $13 - 3 \div 5 =$ ______

S $6 + 8 - 5 =$ ______

R $10 + 5 \div 3 =$ ______

A $7 \times 2 + 5 =$ ______

B $8 \times 3 \div 6 =$ ______

I $18 - 7 \times 0 =$ ______

Find each answer in the code.
Write the letter of the fish above it.

___ ___   ___ ___ ___ ___ ___   ___ ___ ___ ___ ___

0 12   5 0 16 8 5   4 19 12 2 9

Name ______________________

What Can't You Eat for Lunch?

Estimate by rounding each number to the nearer ten.

E $62 + 38$

O $29 + 52$

S $39 + 53$

A $96 + 28$

E $42 + 57$

R $21 + 93$

Estimate by rounding each number to the nearer hundred.

K $422 + 683$

U $927 + 463$

B $701 + 159$

T $481 + 252$

F $395 + 239$

P $719 + 322$

Match the letter of the problem with its estimated answer in the code.

___	___	___	___	___	___	___	___	___
900	110	100	130	1,100	600	130	90	800

___	___		___	___	___	___	___	___
80	110		90	1,400	1,000	1,000	100	110

Name ____________________

Why Did the Leopard Wear a Mask?

Estimate each difference.
Then match the letter of the problem with its answer in the code.

N	$\begin{array}{r} 33 \\ -\ 14 \\ \hline \end{array}$ → $\begin{array}{r} 30 \\ 10 \\ \hline 20 \end{array}$	H	$\begin{array}{r} 72 \\ -\ 39 \\ \hline \end{array}$	A	$\begin{array}{r} 89 \\ -\ 22 \\ \hline \end{array}$
E	$\begin{array}{r} 53 \\ -\ 14 \\ \hline \end{array}$	W	$\begin{array}{r} 85 \\ -\ 26 \\ \hline \end{array}$	D	$\begin{array}{r} 94 \\ -\ 77 \\ \hline \end{array}$
B	$\begin{array}{r} 87 \\ -\ 36 \\ \hline \end{array}$	P	$\begin{array}{r} 593 \\ -\ 246 \\ \hline \end{array}$	I	$\begin{array}{r} 734 \\ -\ 582 \\ \hline \end{array}$
T	$\begin{array}{r} 365 \\ -\ 249 \\ \hline \end{array}$	O	$\begin{array}{r} 761 \\ -\ 493 \\ \hline \end{array}$	S	$\begin{array}{r} 816 \\ -\ 297 \\ \hline \end{array}$

___ ___   ___ ___ ___   ___ ___ ___
30 40   10 100 10   20 300 200

___ ___ ___ ___   ___ ___   ___ ___
60 70 20 200   200 300   50 40

___ ___ ___ ___ ___ ___ ___ .
500 400 300 200 200 40 10

Name ______________________________

Why Did Julio Collect Clocks?

Solve each problem. Write the boxed word from the question above its answer in the code.

1. Rafael likes [to] collect stamps from all over the world. He has 375 stamps in his collection. If 215 stamps are from the United States, how many are from other countries? ________ stamps

2. Sachi decided to [save] seashells. During her vacation at the beach she found 156 shells. She left 54 of the shells at the beach because they were broken. How many shells did she keep? ________ shells

3. Tommy spends a lot of [time] collecting trading cards. He has 623 baseball cards, 111 basketball cards, and 62 football cards. How many trading cards does Tommy have? ________ cards

4. Millie [wanted] to add to her collection of 100 buttons. A friend gave her 32 buttons, and her Aunt Tillie gave her 12 more. How many buttons did Millie have in her collection then? ________ buttons

5. Ina collected 586 coins. She gave away 120 coins [because] they were doubles. How many coins were left in her collection? ________ coins

6. Kwan had 378 miniature cars in his collection. [He] gave 130 away to his best friends Jamal and Henry. How many cars were left in Kwan's collection? ________ cars

________	________	________	________	________	________
466	248	144	160	102	796

Name ______________________

Hidden Picture

Find each sum or difference. Then color each area of the picture below that has one of your answers.

9 + 4	33 + 18	60 − 25	15 − 6	27 + 5
105 + 5	322 − 8	631 − 24	406 + 127	555 − 216

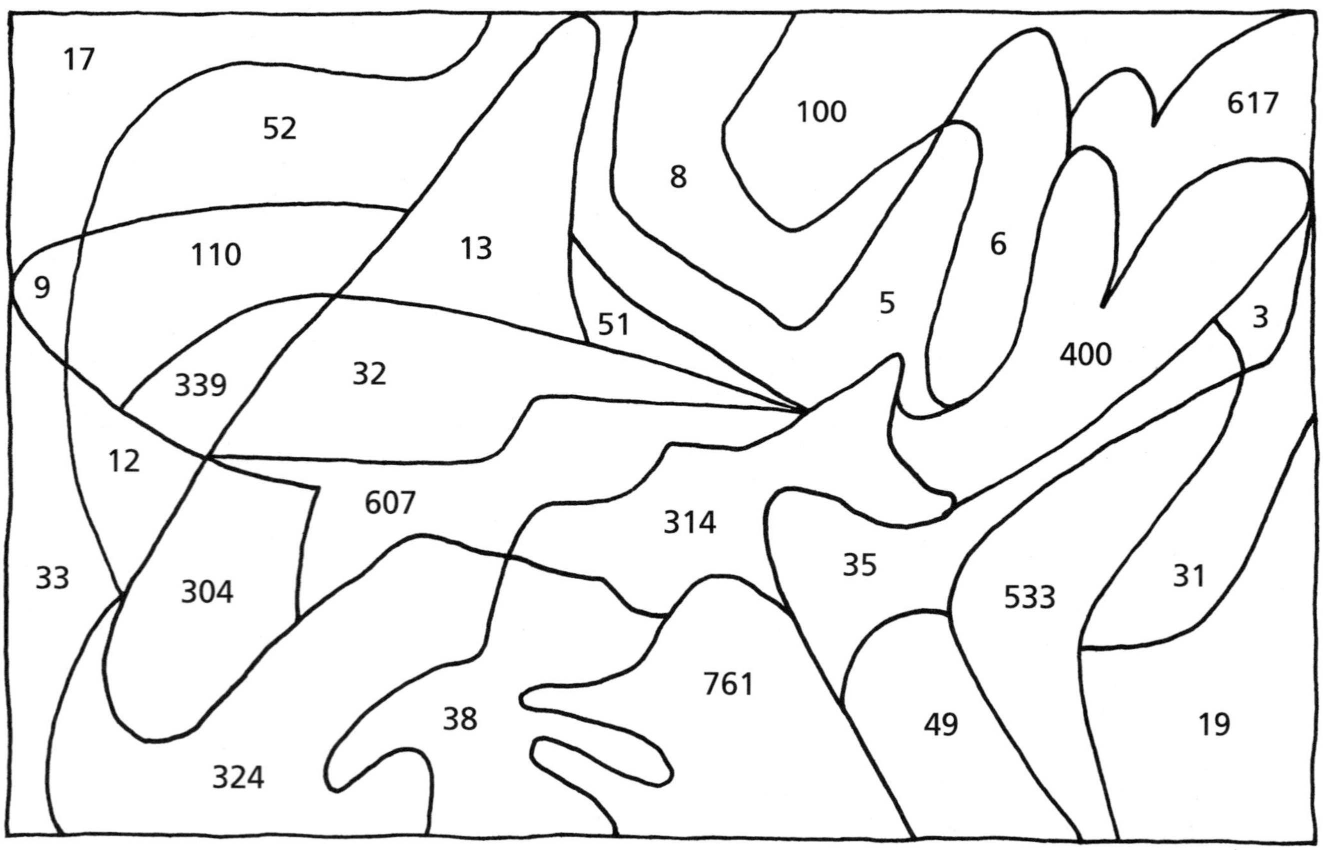

Name ______________________________

What Is a Carnivore's Favorite Game?

Find each sum or difference. Use the answers to complete the cross-number puzzle.

Across

1. $199 + 20$
3. $425 - 75$
5. $325 + 294$
7. $654 + 151$
9. $767 - 484$
10. $839 - 271$
11. $823 - 281$

Down

2. $333 - 152$
4. $820 - 270$
5. $193 + 482$
6. $456 + 472$
7. $273 + 562$
8. $934 - 372$

1.	2.	D		3.	4.	E
	H				L	
5.	W	6.		7.		8.
R	■	9.		A	■	
10.	O			11.	S	T

Locate each number that appears in the code in the cross-number puzzle. You will use only the numbers that have a letter in their box. Write that letter above the number in the code.

__ __ __ __ __ __ __   __ __ __
4 1 3 5 5 6 1   2 8 0

__ __ __ __ __ __
5 0 3 9 0 7

Name ____________________

Why Was the Math Teacher Sad?

Find each sum or difference.
Then write the letter of the exercise above the matching answer at the bottom of the page.

E $23 + 8$	L $70 - 2$	U $119 + 21$	C $345 - 16$
B $100 - 90$	H $345 + 72$	R $304 - 130$	A $119 + 190$
P $222 + 88$	S $321 - 43$	Y $456 + 44$	N $444 + 157$
O $712 - 323$	M $323 - 246$	D $543 + 288$	T $148 - 49$

__ __ __ __ __ __ __   __ __ __
10 31 329 309 140 278 31   278 417 31

__ __ __   __ __   __ __ __ __
417 309 831   278 389   77 309 601 500

__ __ __ __ __ __ __ __
310 174 389 10 68 31 77 278

Name ______________________

Lost in a Cave

Help Travis find his way through the maze.
Follow the path of answers for each multiplication or division fact.

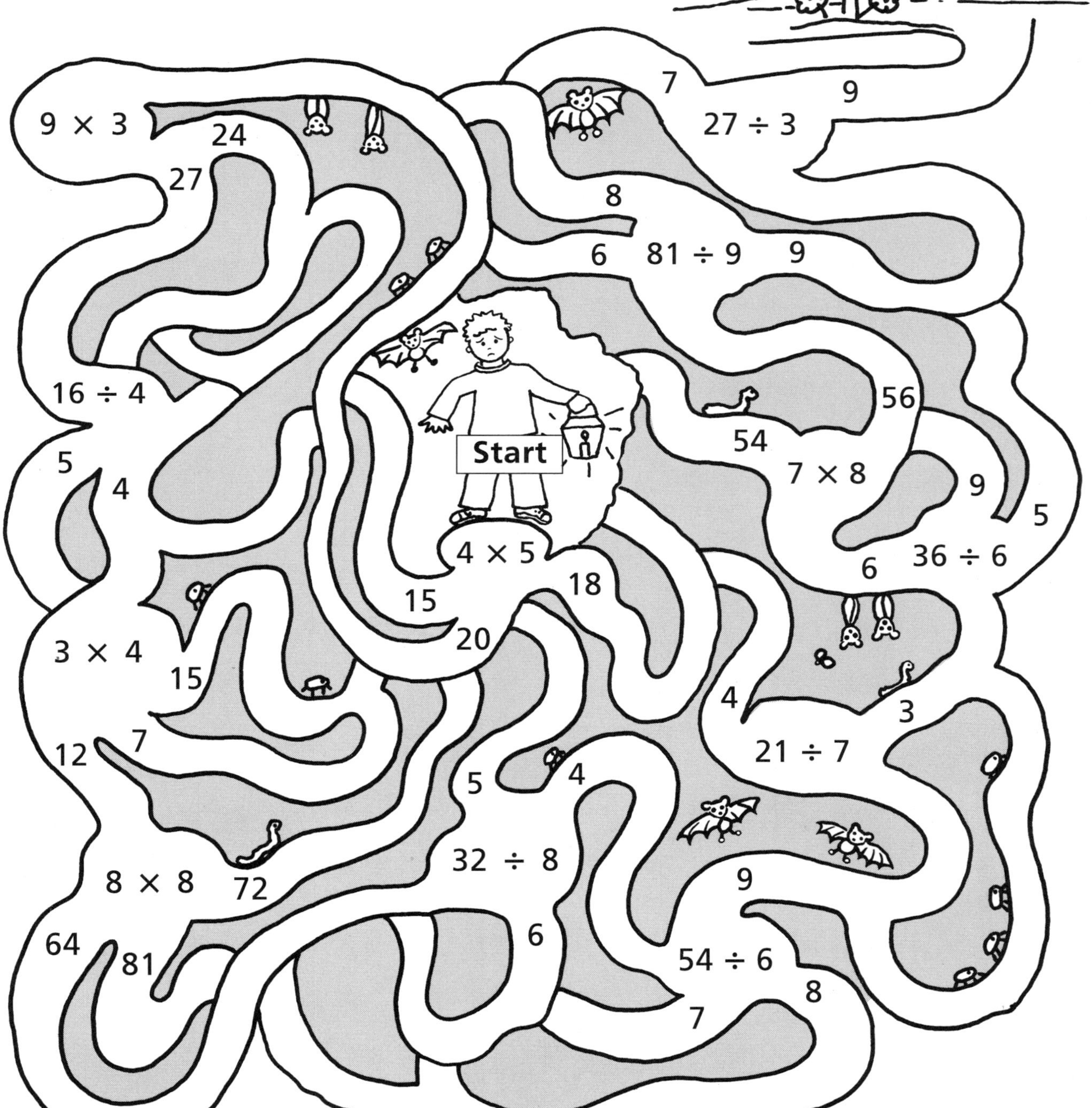

Name ________________________

Digging for Answers

Find each quotient and remainder. Use a ruler and draw lines to match answers that are the same. Your lines will cross out letters. Write the letters that remain in order at the bottom of the page.

37 ÷ 6 = ______ • • 46 ÷ 5 = ______

B T A

14 ÷ 4 = ______ • K E Y G • 13 ÷ 3 = ______

R A W E H

28 ÷ 5 = ______ • U O • 55 ÷ 9 = ______

F A Y

7 ÷ 3 = ______ • C M • 5 ÷ 2 = ______

L R H

9 ÷ 2 = ______ • I S • 70 ÷ 9 = ______

T Q P

47 ÷ 7 = ______ • S H I O • 23 ÷ 7 = ______

B V E

82 ÷ 9 = ______ • • 33 ÷ 6 = ______

T E D

63 ÷ 8 = ______ • L • 53 ÷ 8 = ______

How do you make a dog stop digging?

____.

► Multiply 2 digits by 1 digit without regrouping

Name ____________________

Soaring with Multiplication

Find each product. Then connect your answers, in order, to complete the dot-to-dot.

1. 23×3
2. 11×6
3. 20×1
4. 45×1
5. 32×2
6. 13×2
7. 31×3
8. 28×1
9. 43×2
10. 10×4
11. 20×4
12. 11×5
13. 32×3
14. 25×1
15. 11×7
16. 30×3
17. 68×1
18. 42×2

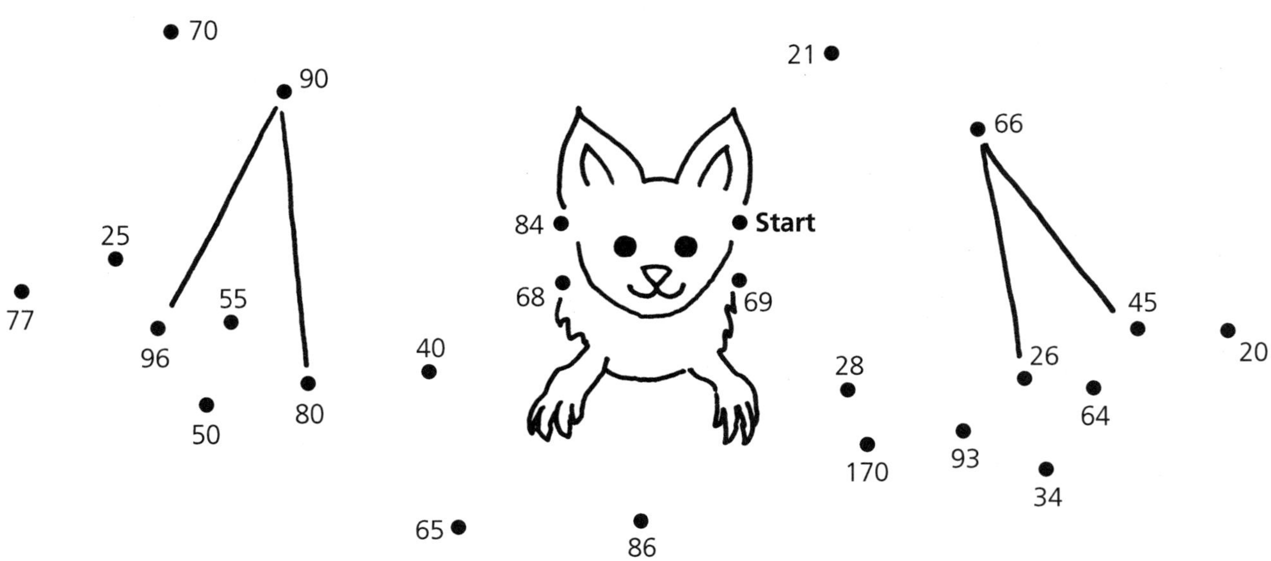

Name ______________________

What a Day!

Find each product. Look for your answers in the story at the bottom of the page. Write the word of the problem in the space above the answer.

Julius	books	morning	forgot	when
25 × 3	32 × 7	18 × 4	16 × 5	39 × 2

home	sneezed	afternoon	finally	bus
13 × 4	37 × 2	43 × 9	22 × 5	49 × 2

day	Cara	homework	bell	knee
56 × 2	68 × 3	38 × 7	28 × 5	39 × 8

bad	playground	lunch	things	tooth
45 × 2	27 × 5	16 × 4	19 × 5	53 × 5

This ______ (72) when I left for school, I ______ (80) my ______ (64).

Then ______ (95) really got ______ (90). I lost my ______ (266).

I scraped my ______ (312) on the ______ (135). ______ (75) ______ (74) on

me. And ______ (78) it was ______ (110) time to go ______ (52),

I missed the ______ (98). I had a really bad ______ (112)!

Name ______________________

Mystery Picture Puzzle

Solve the division problems in the boxes at the bottom of the page. Then carefully cut out the picture pieces. Paste each piece on the box with the matching answer.

9R2	8R2	4R3	5R5
7R2	7R1	6R1	8R4

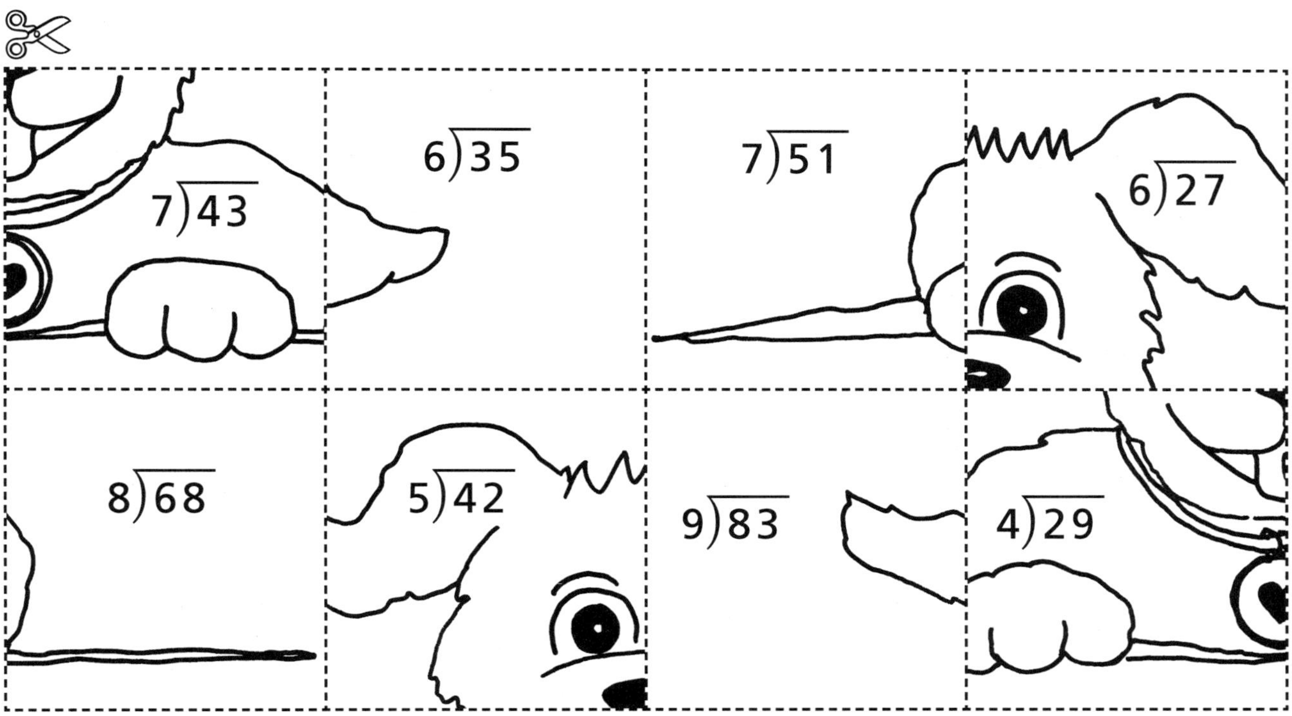

Name ______________________

Why Don't Lobsters and Crabs Share?

Divide. Find your answer in one of the boxes at the bottom of the page and cross out the box. Write the words that remain in order.

1. $8\overline{)62}$
2. $9\overline{)89}$
3. $7\overline{)38}$
4. $9\overline{)38}$
5. $4\overline{)21}$
6. $5\overline{)42}$
7. $8\overline{)33}$
8. $6\overline{)57}$
9. $9\overline{)75}$
10. $4\overline{)26}$
11. $9\overline{)47}$

BECAUSE 7 R6	THEY 5 R3	BOTH 9 R1	ATE 8 R3
ALL 9 R8	OF 6 R2	ARE 7 R4	THEM 5 R1
WANTED 4 R2	JUST 6 R5	PLAIN 3 R2	TO 9 R3
BE 8 R2	SHELLFISH 7 R5	ALONE 4 R1	CLAWS 5 R2

____.

▸ Use mental computation

Name ________________________________

A Riddle for You

Find each answer.

Try to solve each exercise in your head.

M The number of days in 2 weeks ________

E Double 5 and add 7. ________

S Start with 8. Add 3 and subtract 4. ________

L The number of tires on 2 cars ________

O $2 \times 2 \times 2 \times 2 =$ ________

I Start with 4. Add 6 and subtract 5. ________

N The number of fingers on 3 hands ________

E Triple 3 and add 3. ________

T $8 - 2 - 2 - 2 - 2 =$ ________

G Start with 20. Divide by 2 and add 1. ________

O The number of ears on 10 dogs ________

A Double 8 and add it to 8. ________

N $36 \div 6 \div 6 =$ ________

What did the beach say when the tide came in?

Find each answer in the code. Write the letter of the exercise above it.

___ ___ ___ ___   ___ ___ ___ ___   ___ ___   ___ ___ ___

8 20 15 11   0 5 14 17   1 16   7 12 24

Name ______________________________

Why Did the Cookie Go to the Doctor?

Part A Ring the letter of the operation you should use to solve the problem.

1. There are 8 boxes of crayons with 8 crayons in each box. How many crayons in all?

C Multiplication **T** Subtraction **E** Division

2. Rob had 19 balloons. Three balloons popped. How many were left?

A Addition **Y** Subtraction **C** Multiplication

3. Hannah had 35¢. She found a dime. How much money did she have then?

F Subtraction **W** Division **I** Addition

4. There are 6 dogs and 48 dog biscuits. If shared equally, how many biscuits will each dog get?

N Addition **O** Multiplication **R** Division

5. There are 5 shelves with 10 books on each shelf. How many books in all?

P Division **F** Multiplication **S** Subtraction

Part B Ring the letter of the correct answer.

6. The number of inches in 2 feet	**L** 24	**E** 2	**R** 36
7. Triple 10 and then subtract 5	**D** 30	**P** 20	**M** 25
8. How many horns do 3 bulls have?	**A** 9	**E** 6	**I** 4
9. The number of thumbs on 5 hands	**Z** 8	**I** 6	**U** 5
10. Double 50 and subtract 25	**F** 125	**T** 75	**X** 50

Match each letter with a ring to the number of the exercise.

___ ___ ___ ___ ___ ___ ___ ___ ___ ___ ___ ___.
3 10 5 8 6 10 1 4 9 7 7 2

Name ____________________

Nursery Rhyme Riddle

Estimate the cost of each purchase. Ring the letter that shows whether the child has **not enough** or **more than enough** money.

Paint box	$2.49	Eraser	$.90
Crayons	$1.99	Markers	$3.99
Glue	$.99	Scissors	$3.19
Notebook	$2.29	Paintbrushes	3 for $2.79
Folder	$.20	Pencils	2 for $.75
Ruler	$.59		

1. Judy needs to buy 2 pencils and a bottle of glue. She has $2.00.
P not enough **C** more than enough

2. Mike has $7.00. He needs 1 package of markers, scissors, and 2 folders.
T not enough **D** more than enough

3. Alita wants 1 notebook, a ruler, and a box of crayons. She has $4.00.
U not enough **B** more than enough

4. Wayne needs a paint box, 3 paintbrushes, and a notebook. He has $8.00.
A not enough **O** more than enough

5. Darius would like to buy 2 folders, 2 pencils, and a ruler. He has $1.50.
R not enough **S** more than enough

6. Modupé has $4.00 to buy an eraser, 4 pencils, and a bottle of glue.
E not enough **I** more than enough

7. Omar has $5.00. He needs a notebook, 3 folders, 2 pencils, and an eraser.
N not enough **M** more than enough

What did Old Mother Hubbard say when she opened her cupboard?
Write the letter with a ring above the number of the exercise.

___ ___ ___ ___ ___ ___ ___
4 6 1 3 5 7 2

Name ______________________

Check the Estimate

Decide if each estimated answer is an **overestimate** or an **underestimate**.
Color overestimates red. Color underestimates blue.

- \$37.43 + 9.75 = \$50.00
- 620 − 295 = 400
- 383 + 508 = 1,000
- 73 + 442 + 21 = 500
- 896 − 203 = 600
- 28 + 19 = 60
- \$8.79 − 4.20 = \$4.00
- \$1.75 + 4.68 = \$6.00
- 78 + 32 + 25 = 150
- \$60.00 − 29.95 = \$40.00
- 103 + 79 + 22 = 180
- 945 − 580 = 200
- 73 + 18 = 100
- \$1.29 + 3.37 = \$4.00
- 82 − 57 = 30
- 510 − 295 = 300
- 563 + 219 = 700
- \$9.25 + 7.60 = \$18.00
- \$21.50 − 8.62 = \$14.00

Name ______________________

Those A-Maze-ing Fractions

Follow the path of fractions that name the part of each figure that is **not** shaded.

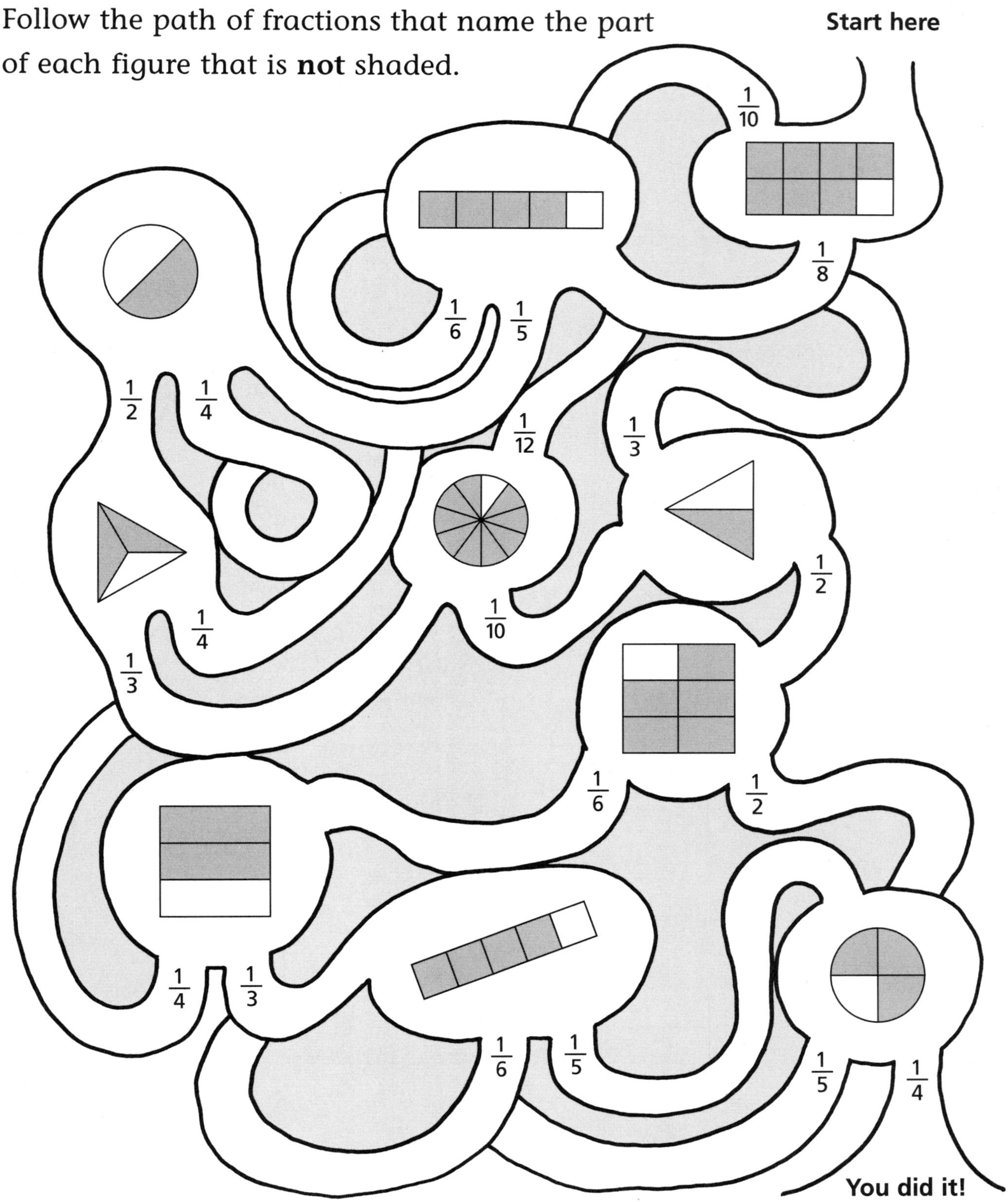

Name ______________________________

What Can You Break with Your Voice?

Follow the directions in each box. Then match the letter of the exercise with its answer in the code.

C Shade $\frac{3}{8}$

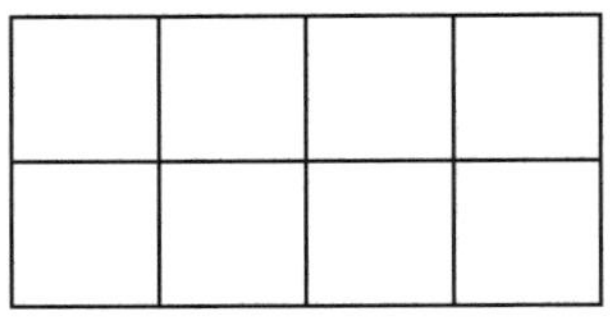

Write the fraction that is not shaded. ________

I Shade $\frac{2}{5}$

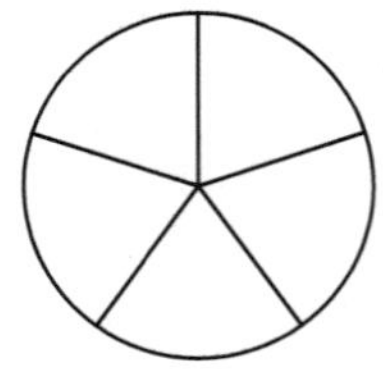

Write the fraction that is not shaded. ________

L Shade $\frac{5}{6}$

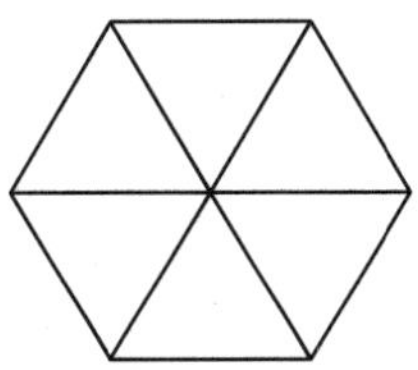

Write the fraction that is not shaded. ________

S Shade $\frac{2}{3}$

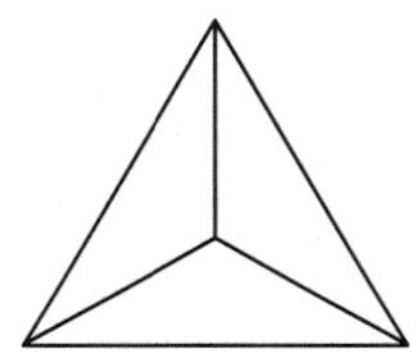

Write the fraction that is not shaded. ________

E Shade $\frac{3}{10}$

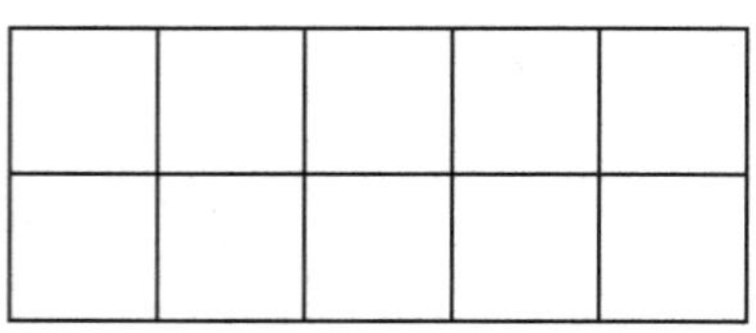

Write the fraction that is not shaded. ________

N Shade $\frac{5}{8}$

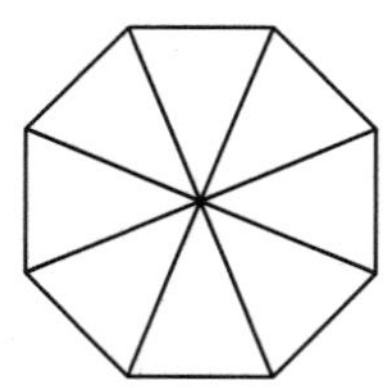

Write the fraction that is not shaded. ________

___	___	___	___	___	___	___
$\frac{1}{3}$	$\frac{3}{5}$	$\frac{1}{6}$	$\frac{7}{10}$	$\frac{3}{8}$	$\frac{5}{8}$	$\frac{7}{10}$

Name ______________________________

What Rooms Have No Windows?

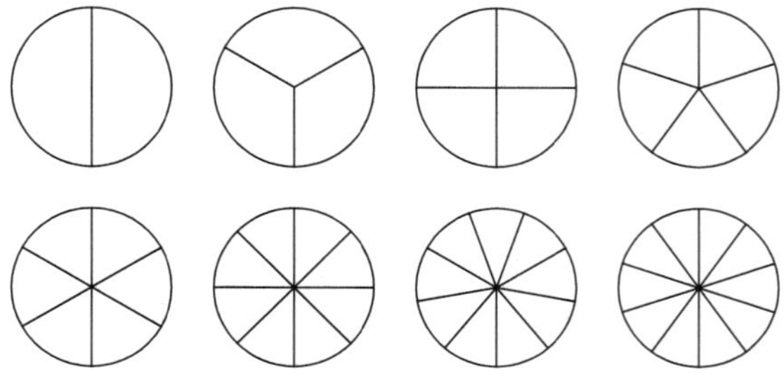

Compare the fractions. Write >, <, or = in the space between each pair.

1. $\frac{1}{3}$ **S** $\frac{1}{4}$ **R**

2. $\frac{2}{8}$ **O** $\frac{2}{5}$ **M**

3. $\frac{1}{8}$ **T** $\frac{3}{8}$ **R**

4. $\frac{3}{6}$ **E** $\frac{3}{4}$ **H**

5. $\frac{5}{9}$ **U** $\frac{2}{9}$ **A**

6. $\frac{6}{10}$ **M** $\frac{3}{10}$ **N**

7. $\frac{1}{2}$ **G** $\frac{2}{2}$ **S**

8. $\frac{5}{8}$ **W** $\frac{5}{6}$ **O**

Match the letter below the greater amount in each pair with the number of the problem.

___	___	___	___	___	___	___	___	___
2	5	7	4	3	8	8	6	1

Name ______________________

The Best Medicine

Solve. Then match the letter of the exercise with its answer in the code.

Mrs. Izzary baked 2 dozen (24) muffins. She frosted $\frac{1}{2}$ of them. She sprinkled $\frac{1}{4}$ of the muffins with powdered sugar, and she left $\frac{1}{4}$ of the muffins plain.

T ________ muffins were frosted

A ________ muffins had powdered sugar

A ________ muffins were plain

Mrs. Izzary gave $\frac{1}{6}$ of the muffins she made to her neighbor.

B ________ muffins were given to the neighbor

R ________ muffins remained

The Izzarys ate $\frac{1}{4}$ of the remaining muffins for dessert that night.

W ________ muffins were eaten for dessert

L ________ muffins remained

The next morning, Mrs. Izzary packed $\frac{1}{3}$ of the remaining muffins in lunches.

W ________ muffins were packed in lunches

E ________ muffins remained

What should you drink when you are sick?

___	___	___	___		___	___	___	___	___
5	10	15	15		5	6	12	10	20

► Identify mixed numerals

Name ______________________________

Take a Bite Out of Mixed Numerals

Note: Use with page 36.

Write the mixed numeral that describes the part of each group that is shaded.

1. 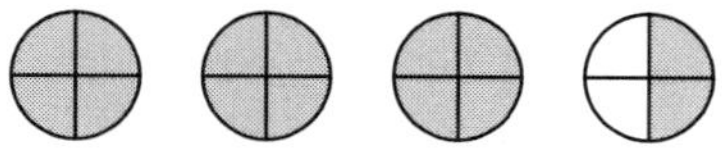 ________ → 

2. 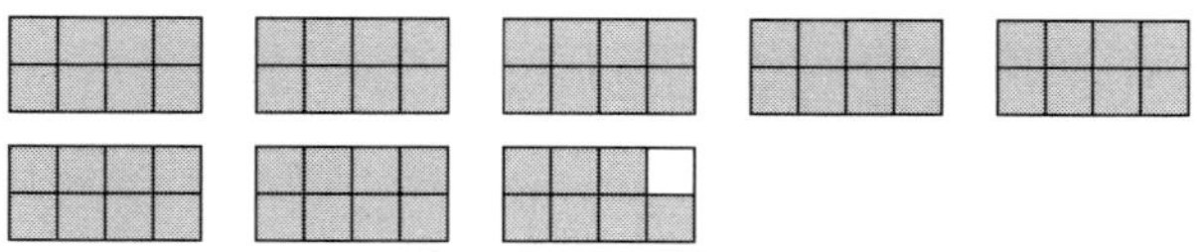 ________ → 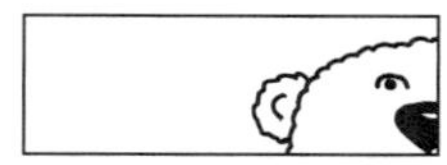

3. 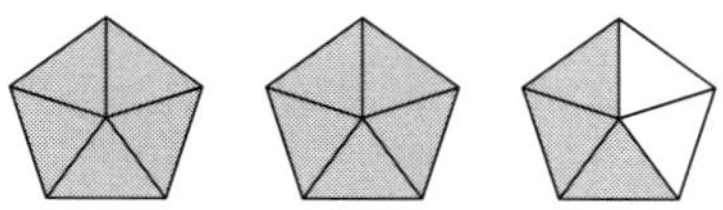 ________ → 

4. 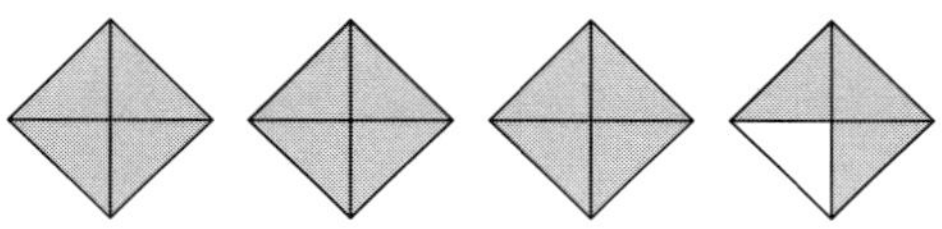 ________ → 

5. 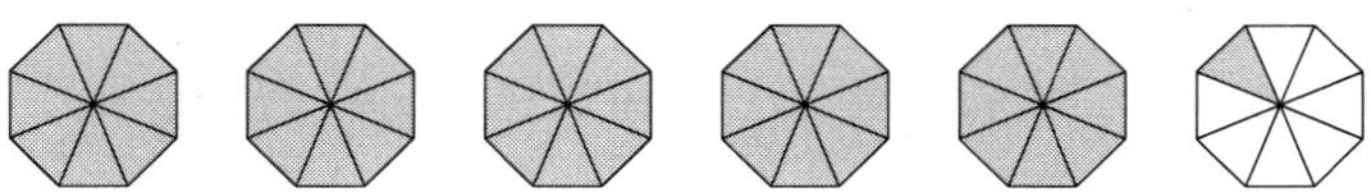 ________ → 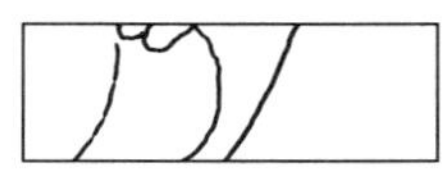

6. 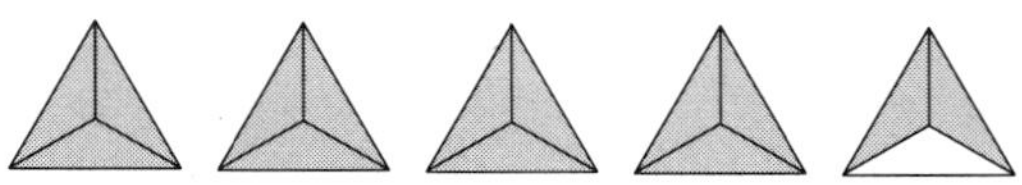 ________ → 

7. 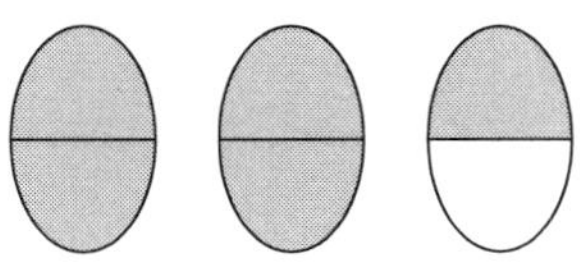 ________ → 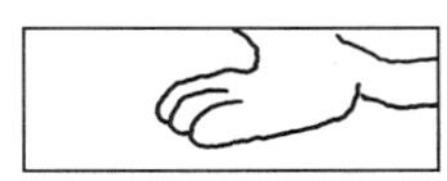

8. 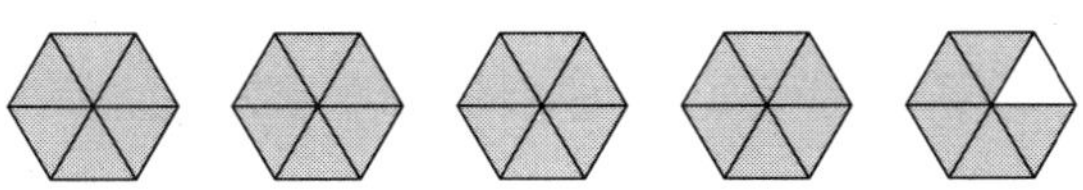 ________ → 

9. 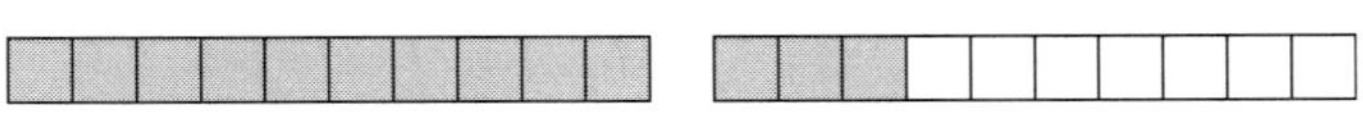 ________ → 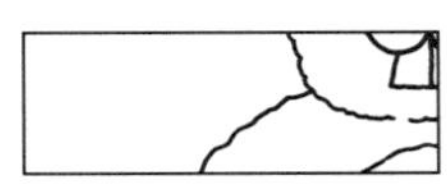

10. 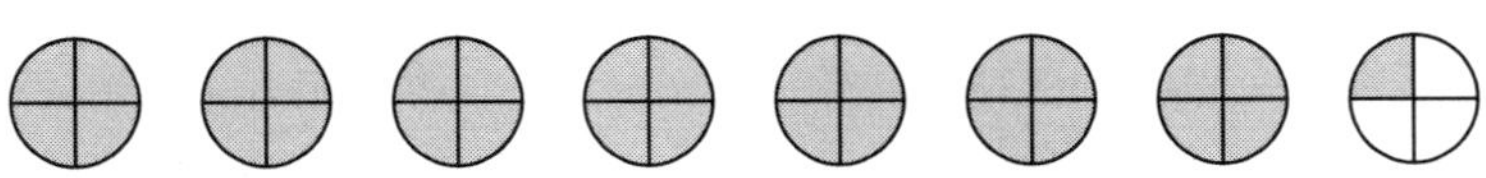 ________ → 

Name ____________________

Take a Bite Out of Mixed Numerals

Note: Use with page 35.

Carefully cut out the picture pieces. Match each picture piece to one shown on page 35 and note the mixed numeral that goes with it. Then paste the piece in the box below that has the same mixed numeral.

$7\frac{7}{8}$	$4\frac{5}{6}$
$1\frac{3}{10}$	$3\frac{1}{2}$
$2\frac{3}{5}$	$7\frac{1}{4}$
$3\frac{3}{4}$	$5\frac{1}{8}$
$2\frac{1}{2}$	$4\frac{2}{3}$

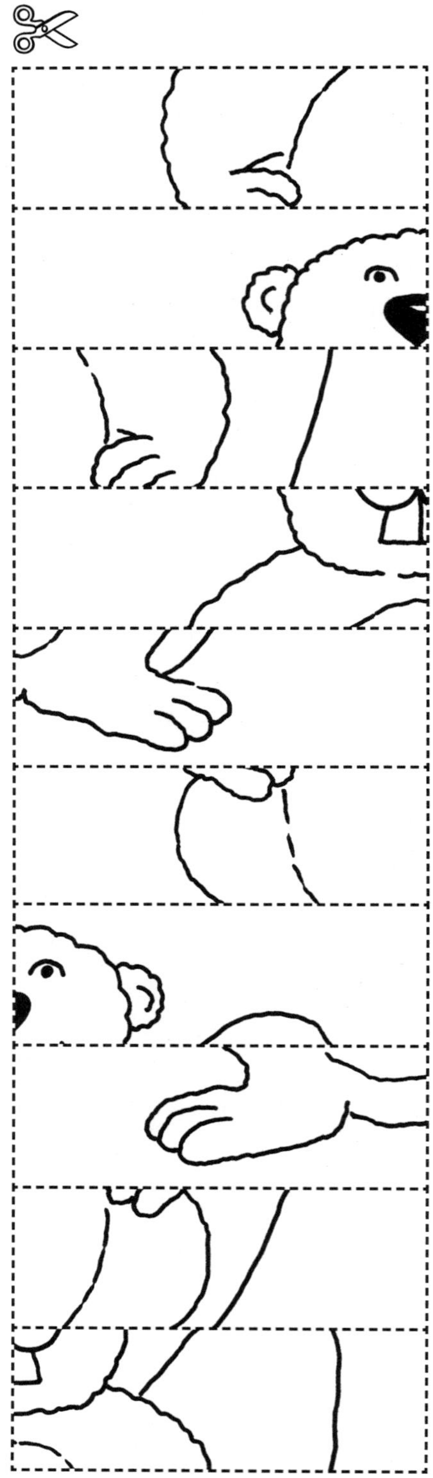

Name ______________________________

When Is the Best Time to See a Dentist?

Ring the letter of the correct answer. Use the calendar to answer questions 1–5.

January						
Sun.	Mon.	Tues.	Wed.	Thurs.	Fri.	Sat.
		1	2	3	4	5
6	7	8	9	10	11	12
13	14	15	16	17	18	19
20	21	22	23	24	25	26
27	28	29	30	31		

1. The 12th of January is
  S Monday **E** Sunday **T** Saturday

2. The date for the 3rd Monday of the month is the
  I 14th **O** 21st **R** 3rd

3. The first day of February is
  O Friday **G** Sunday **M** Tuesday

4. The date for the 1st Sunday of the month is the
  H 1st **T** 6th **A** 5th

5. The month of January ends on a
  T Saturday **E** Friday **H** Thursday

6. The last month of the calendar year is
  H December **D** September **N** July

7. The month right before June is
  F July **B** January **U** May

8. Hans started reading a book in March and finished 3 months later. He finished reading the book in
  R June **T** July **M** May

9. The 8th month of the year is
  Y July **S** September **T** August

10. The 4th month of the year is
  L May **Y** April **N** March

Write the letters of your answers in order.

____

Name ____________________

What Do You Do First Each Day?

Solve.

1. It took Ina 15 minutes to walk to school. If she left at 7:25 a.m., what time did she get to school? __________
2. At 8:05 a.m., Ina's teacher said that a special visitor was supposed to arrive in 40 minutes. What time was the visitor expected? __________
3. The visitor arrived 5 minutes late. He spent 30 minutes showing his collection of pet snakes. What time did he leave? __________
4. The class spent the next 25 minutes writing thank-you notes to the visitor. What time did they finish their thank-you notes? __________
5. Reading time started at 9:45 a.m. It lasted 1 hour and 15 minutes. When did reading time end? __________
6. Art class started at 1:30 p.m. The class was 40 minutes long. What time was art class over? __________
7. After school, Ina went to a friend's house. It took her 25 minutes to walk home from there. She got home at 5:00 p.m. What time did she leave her friend's house? __________
8. Ina has 45 minutes worth of homework. It takes her 20 minutes to get ready for bed. If Ina's bedtime is 9 p.m., what is the latest time she can start her homework and still get to bed on time? __________

Find each answer in the Code Box. Then match the letter in front of the answer to the number of the question in the code.

Code Box

Y	4:35 p.m.	**A**	9:20 a.m.	**G**	8:00 a.m.	**W**	7:55 p.m.	**E**	9:45 a.m.	**R**	2:30 p.m.
I	9:15 a.m.	**U**	7:40 a.m.	**P**	8:45 a.m.	**S**	8:45 p.m.	**K**	11:00 a.m.	**O**	2:10 p.m.

___ ___ ___   ___ ___ ___ ___   ___ ___
7 6 1   8 3 5 4   1 2

Name ______________________________

It Is About Time

Solve.

W Chin took his dog for a walk. They left the house at 6:50 a.m. and returned at 7:40 a.m. How many minutes were they gone? ________

G Kelsey and her friend went to see a movie. The show started at 2:15 p.m. and ended at 3:55 p.m. How long was the movie? ________

N Becca's family started out for the museum at 1:05 p.m. The trip took 1 hour and 10 minutes. What time did they get to the museum? ________

R Rosa baked cookies. The first batch went in the oven at 4:10 p.m. and came out at 4:22 p.m. How long did the cookies bake? ________

U Ethan went shopping with his mother. They left home at 10:30 a.m. and returned at 12:00 noon. How long were they gone? ________

K Randy's guitar lesson began at 4:15 p.m. If the lesson lasted for 30 minutes, at what time was the lesson over? ________

M Sadie, the cat, finished eating 7:20 p.m. She had started eating 25 minutes before. What time was it when Sadie started eating? ________

T Diane finished her science project at 3:00 p.m. She began working at 1:35 p.m. How much time did she spend working? ________

S A pizza was delivered to Bernard's house at 5:25 p.m. It had been ordered at 4:45 p.m. How long did it take for the pizza to arrive? ________

I Tracey was tired and decided to take a nap. She fell asleep at 1:45 p.m. and woke up at 2:55 p.m. How long did Tracey sleep? ________

What do you get when you cross a fish with an elephant?

Write the letter of the question above its answer.

___	___	___	___	___	___	___	___
40 min	50 min	1 hr 10 min	6:55 p.m.	6:55 p.m.	1 hr 10 min	2:15 p.m.	1 hr 40 min

___	___	___	___	___	___
1 hr 25 min	12 min	1 hr 30 min	2:15 p.m.	4:45 p.m.	40 min

Name ______________________

Interesting Temperature Facts

Read each thermometer to find the temperature.

A You can tell the outside temperature by the number of times a cricket chirps. If it chirps 8 times in 15 seconds, the temperature is ________ °F.

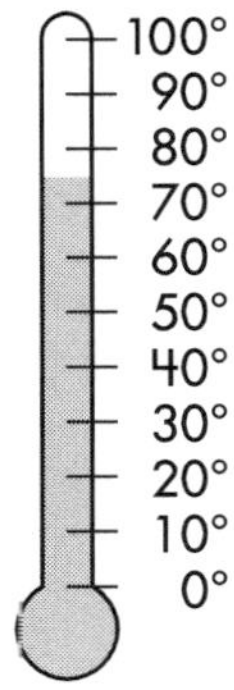

C The hottest city in the United States is Yuma, Arizona. The average high temperature there is ________ °F.

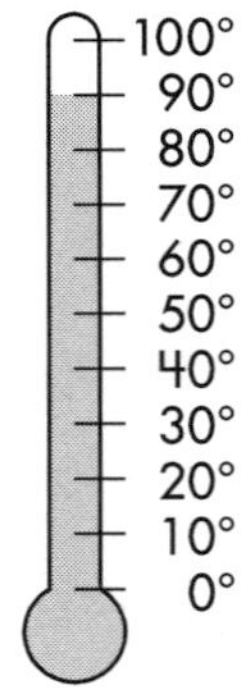

O The temperature of a sea turtle's egg tells whether it will hatch as a boy or girl turtle. It will be a girl if the temperature is ________ °F.

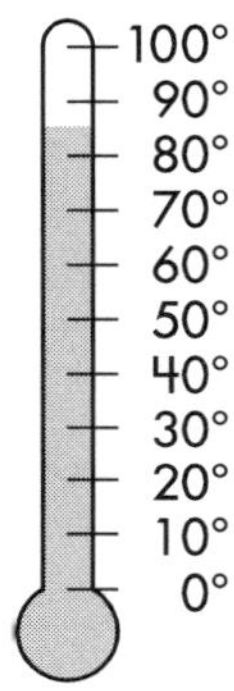

E At the zoo in Phoenix, animals are given ice pops made of frozen fruits, vegetables, fish, and seeds to help keep the animals cool in Phoenix's usual summer temperature of ________ °F.

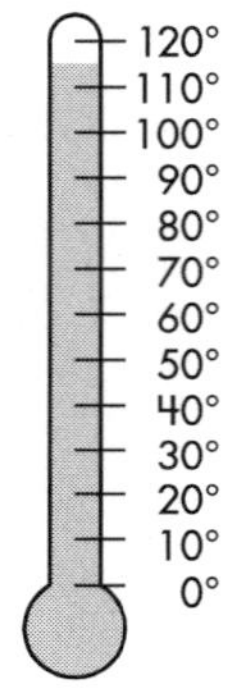

L Baby chicks must be kept very warm when they hatch. The best temperature to keep them healthy is ________ °F.

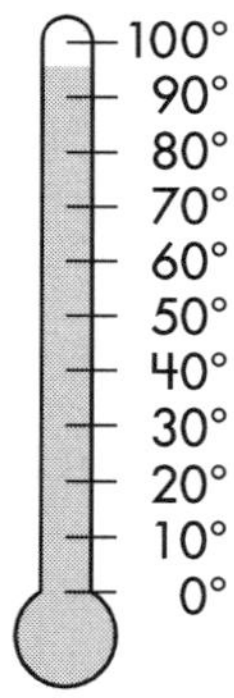

R Caves stay at about the same temperature all year. The temperature in most caves is around ________ °F.

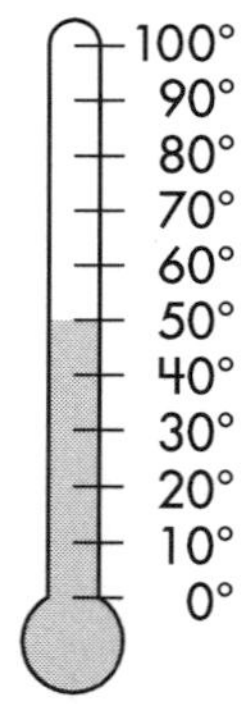

Write the letter of the exercise above its answer.

___	___	___	___		___	___	___	___
50°	115°	75°	95°		90°	85°	85°	95°

Name ______________________________

Measure Treasure

Travel through the maze by choosing the better unit to measure each thing described.

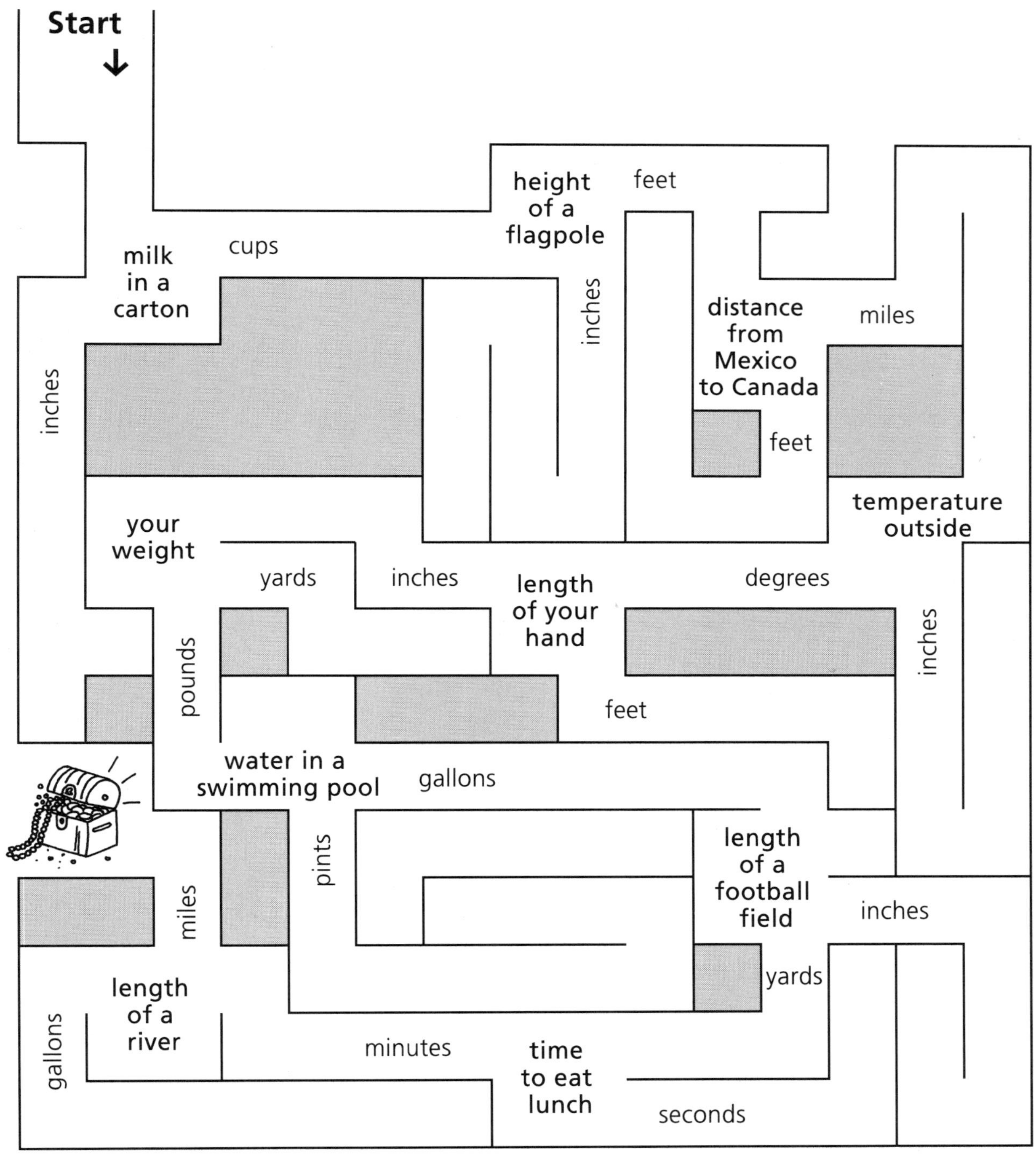

Name ______________________

An Interesting Fact

Ring the letter of the better metric unit to use to measure each thing described.

1.	Length of a playground	**F**	meters	**D**	liters
2.	Amount of water in a glass	**P**	millimeters	**O**	milliliters
3.	Weight of a cupcake	**W**	centimeters	**H**	grams
4.	Length of a pencil	**E**	centimeters	**L**	kilometers
5.	Weight of a watermelon	**A**	kilograms	**C**	grams
6.	Amount of water in a fish tank	**V**	centiliters	**T**	liters
7.	Distance from Japan to China	**R**	kilometers	**B**	centimeters
8.	Height of a flagpole	**G**	centimeters	**S**	meters
9.	Thickness of a dime	**Y**	meters	**J**	millimeters
10.	Weight of a paper clip	**M**	grams	**U**	kilograms
11.	Amount of medicine in an eyedropper	**N**	milliliters	**K**	milligrams

Match the letter of the answer to the number of the question in the code. When you are finished, you will find out the name of the U.S. President who was the first to want the United States to adopt the metric system.

___	___	___	___	___	___
6	3	2	10	5	8

___	___	___	___	___	___	___	___	___
9	4	1	1	4	7	8	2	11

Name ______________________

What Wears Shoes but Has No Feet?

Ring the letter of the better estimate.

1.

P 1 gram **E** 1 kilogram

2.

L 10 grams **M** 10 kilograms

3.

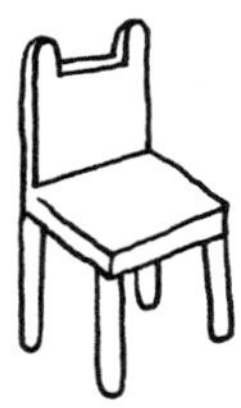

A 12 grams **D** 12 kilograms

4.

A 1 gram **I** 1 kilogram

5.

K 30 grams **I** 30 kilograms

6.

R 50 grams **W** 50 kilograms

7.

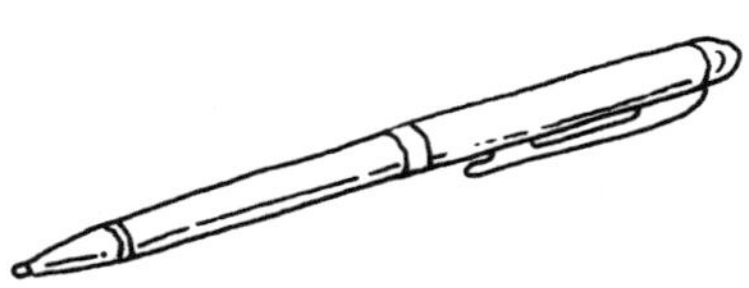

S 7 grams **T** 7 kilograms

8.

O 6 grams **K** 6 kilograms

Match the letter with a ring with the number of its box.

___   ___ ___ ___ ___ ___ ___ ___ ___
4   7 5 3 1 6 4 2 8

▸ Measure length in customary units

Name ____________________

What Has One Foot on Each Side and One in the Middle?

Measure each line to the nearer inch. Then write the letter of the line above your measurement in the code at the bottom of the page.

T ____________________ K ____

D ______ R ____

S ____

I __________________________________ Y ________________________

C ____

A ____

____   ____ ____ ____ ____ ____ ____ ____ ____ ____
8 in.   3 in. 8 in. 6 in. 1 in. 7 in. 2 in. 4 in. 9 in. 5 in.

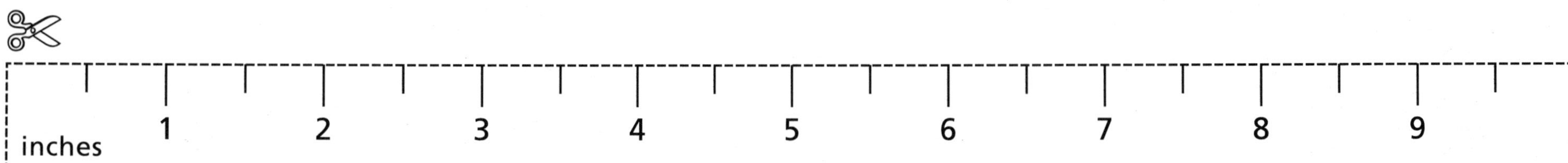

Name ____________________

How Can You Recognize a Dogwood Tree?

Measure each line to the nearer centimeter. Then write the letter of the line above your measurement in the code at the bottom of the page.

A ______ K ________________ S ____

Y ______________________________

B ________________________ T ________

B __________________________________

R ______________________ I ____________________

____ 16 cm ____ 15 cm ____ 9 cm ____ 4 cm ____ 2 cm ____ 12 cm ____ 3 cm ____ 11 cm ____ 8 cm

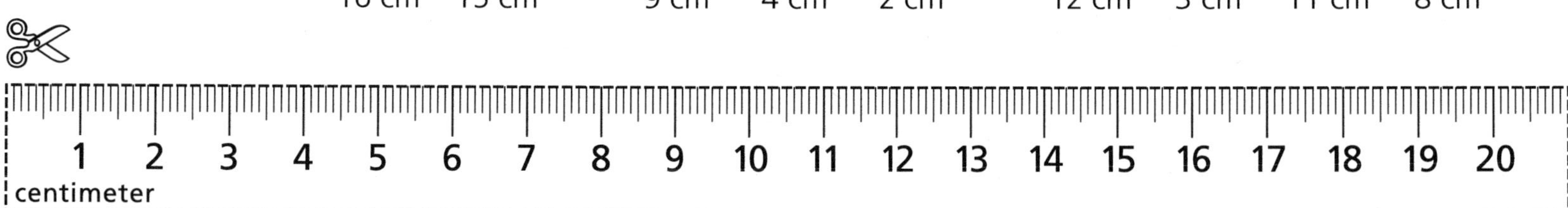

Name ____________________

Around You Go!

Find the perimeter of each figure. Then shade the part of the design that has your answer.

1.

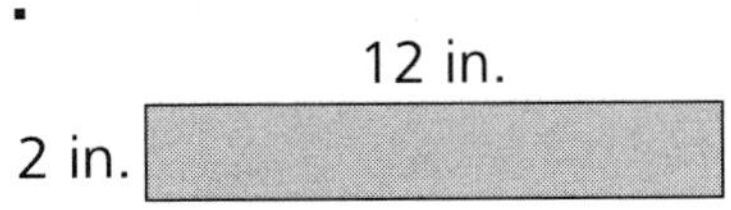

____

2.

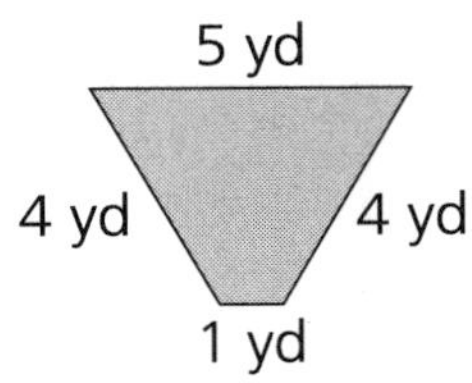

____

3.

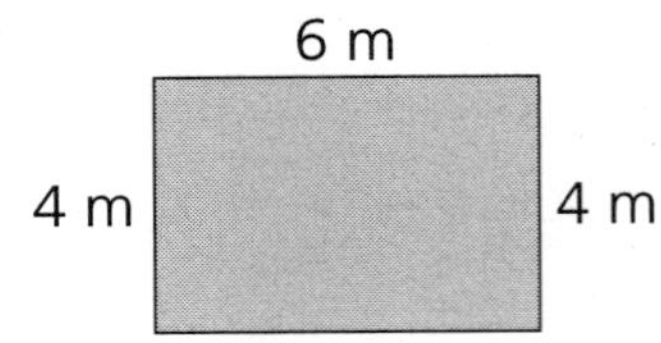

____

4.

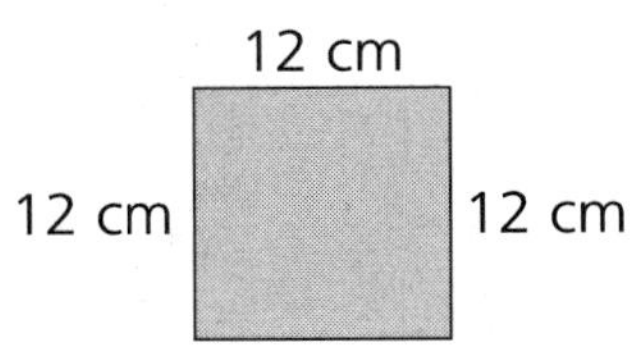

____

5.

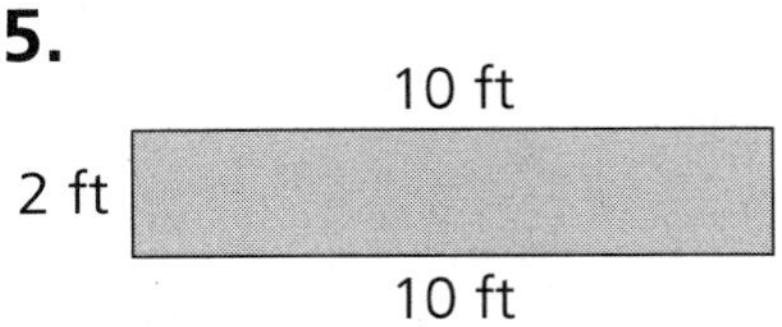

____

6.

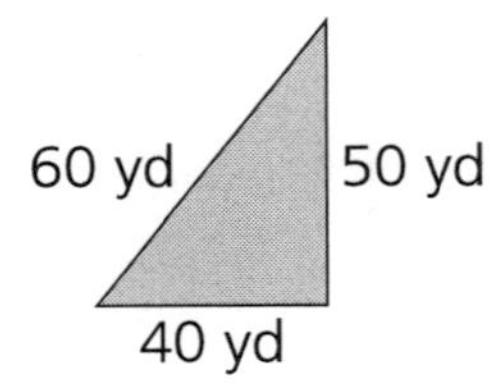

____

7.

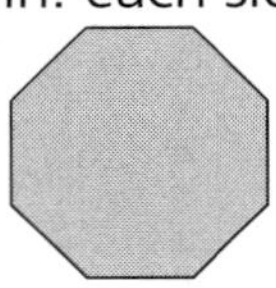

____

8.

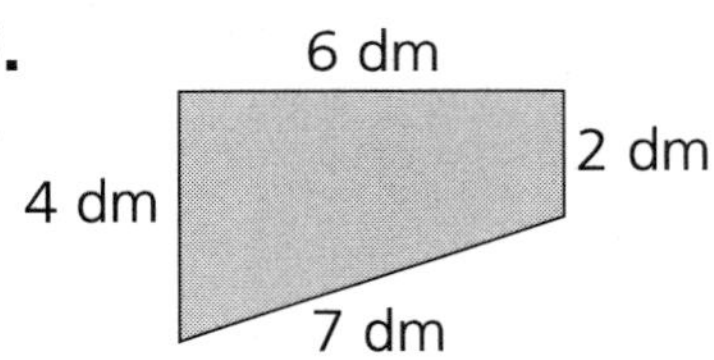

____

9.

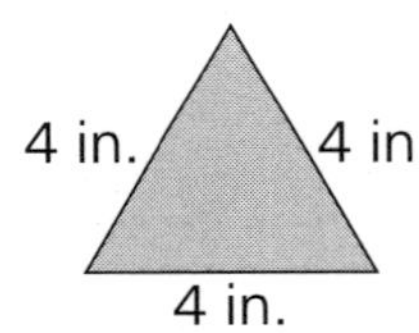

____

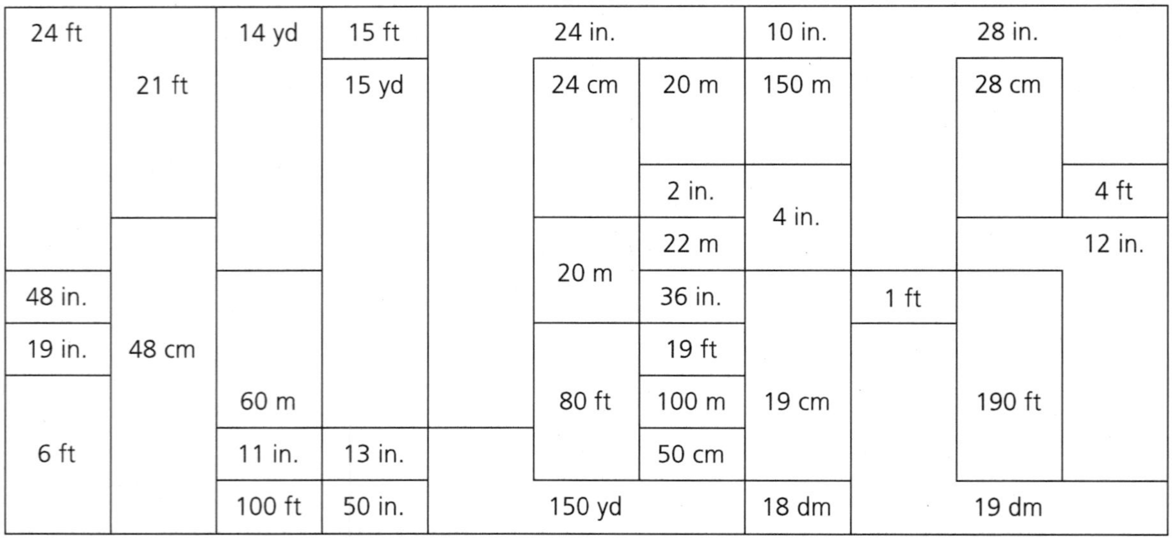

Name ______________________

Different but Equal

Find the area of each shaded part. Use a ruler and draw lines to match two areas that are the same. Your lines will cross out letters. Write the letters that remain in order at the bottom of the page.

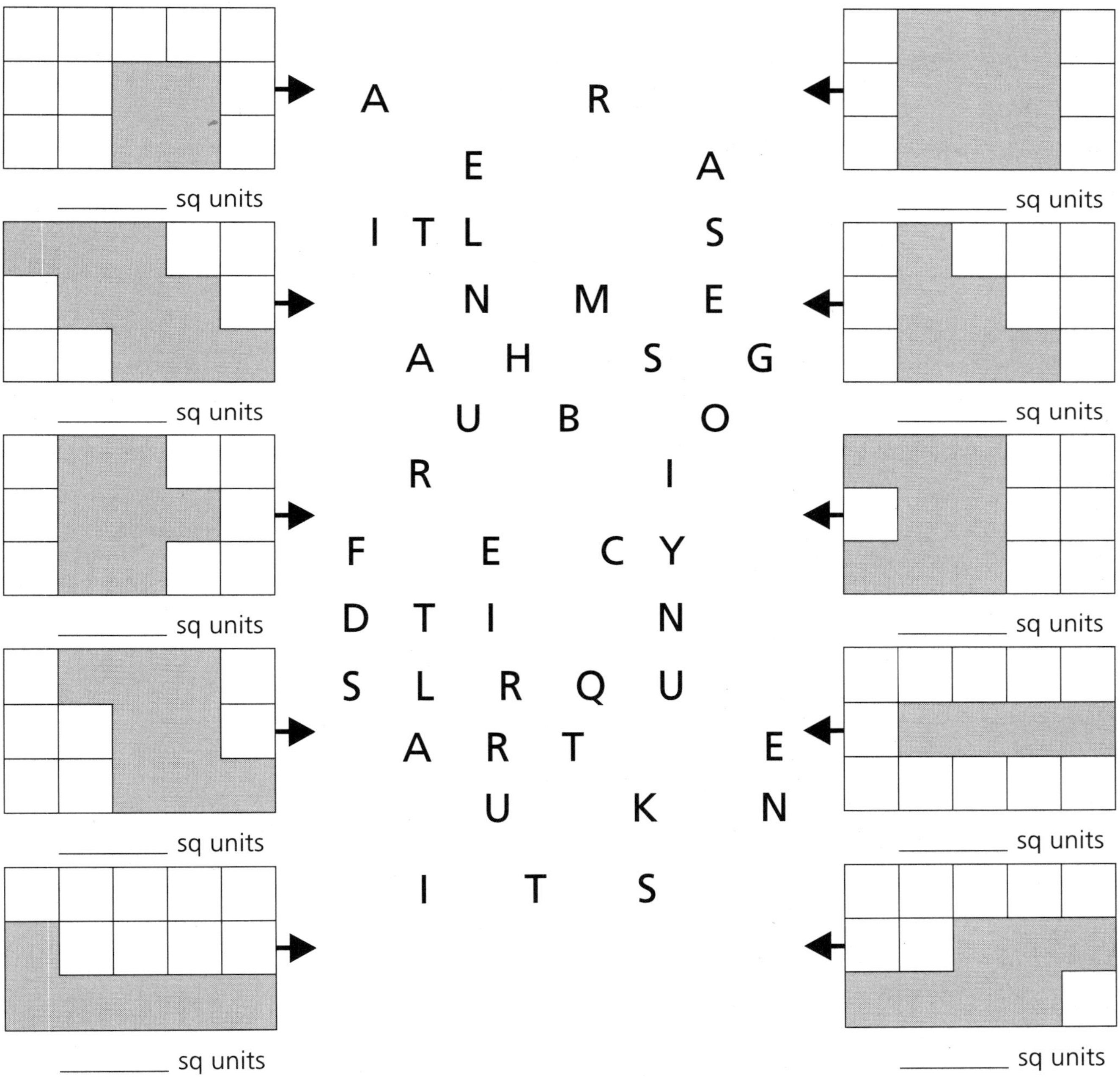

____

Name ______________________

What Did the Guitar Say to the Rock Star?

Find the perimeter and area of each figure.

Example: □ perimeter = 4 units, area = 1 sq unit

1.

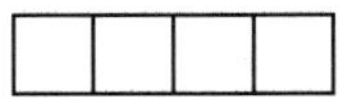

perimeter = ________ units → **I**

area = ________ sq units → **K**

2.

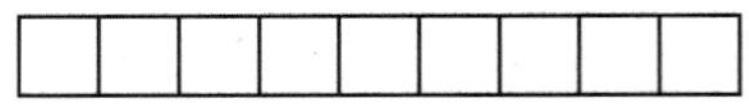

perimeter = ________ units → **M**

area = ________ sq units → **G**

3.

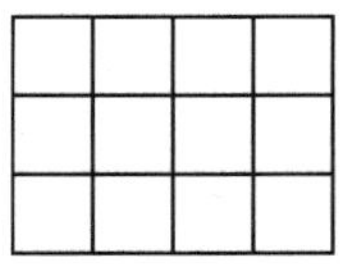

perimeter = ________ units → **O**

area = ________ sq units → **E**

4.

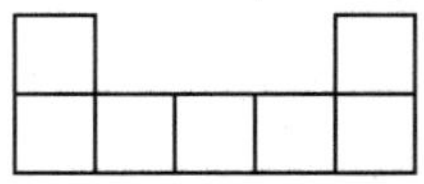

perimeter = ________ units → **C**

area = ________ sq units → **T**

5.

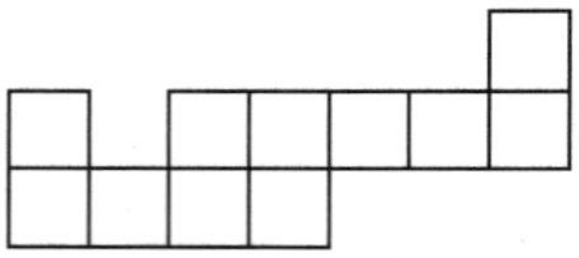

perimeter = ________ units → **P**

area = ________ sq units → **U**

6.

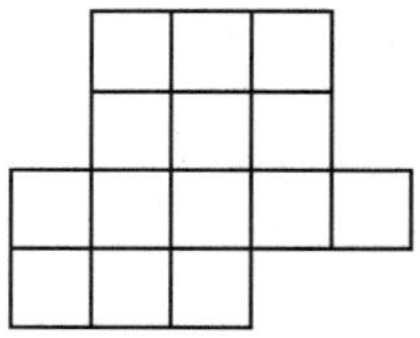

perimeter = ________ units → **Q**

area = ________ sq units → **N**

Match the letter with the measurement.

____	____	____	____
18 units	11 sq units	10 units	7 sq units

____	____	____	____	____	____	____
22 units	10 units	16 units	4 sq units	10 units	14 sq units	9 sq units

____	____		____	____ .
14 units	14 sq units		20 units	12 sq units

Name ______________________

What Are They?

Write the number that tells how many cubes are used to make each figure. Use a ruler and draw lines to match two figures that have the same number of cubes. Your lines will cross out letters. Write the letters that remain in order at the bottom of the page.

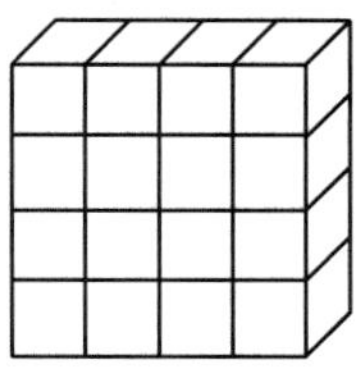

____ cubes →

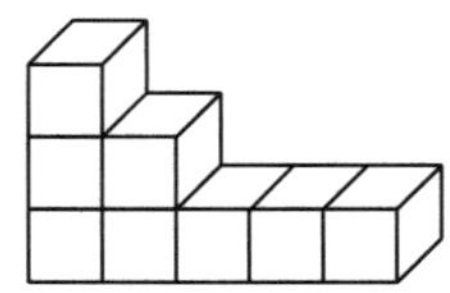

← ____ cubes

T H E

I

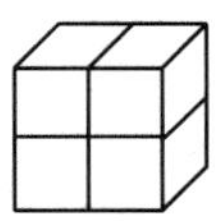

____ cubes →

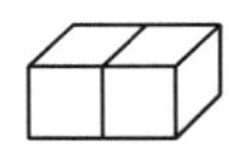

← ____ cubes

J O B

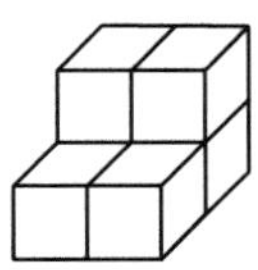

L M

____ cubes →

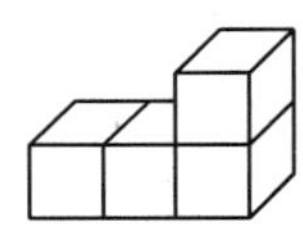

← ____ cubes

D

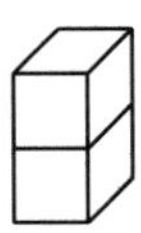

P I E

____ cubes →

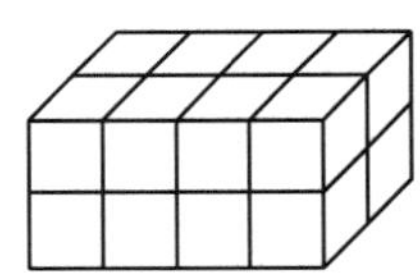

← ____ cubes

K

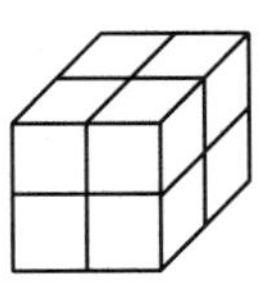

A S G

____ cubes →

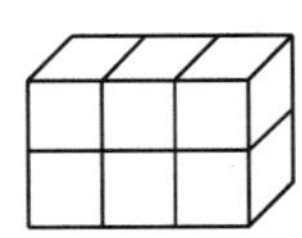

← ____ cubes

The more you take away from them, the bigger they get. What are they?

____

Name ______________________

What Do All 3-D Figures Have?

Which solid in each row can be made from the net shown?
Ring the letter below that solid.

1. 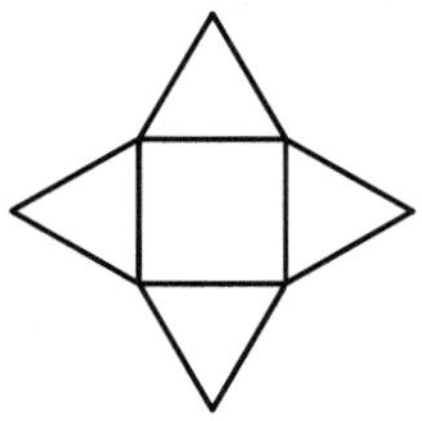	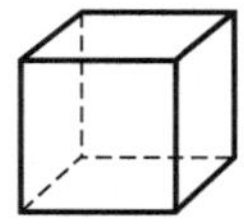 P	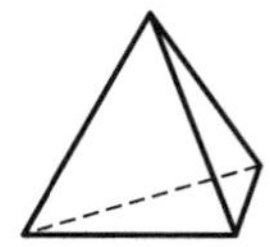 R	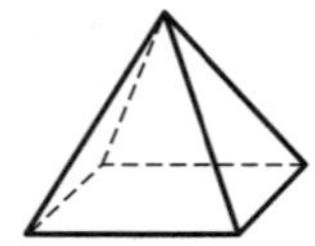 S
2. 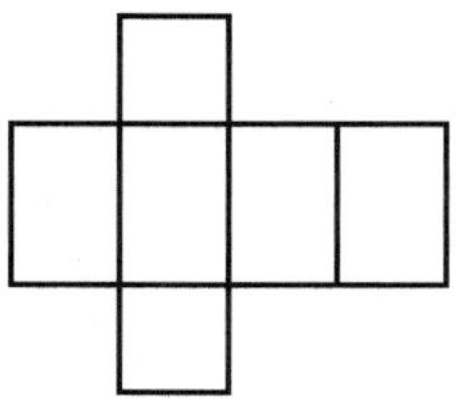	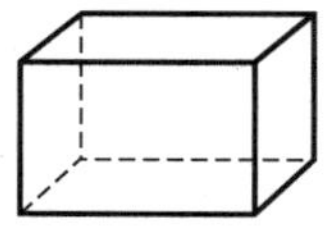 A	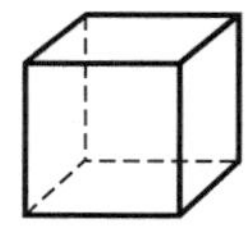 K	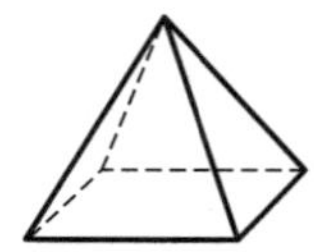 I
3. 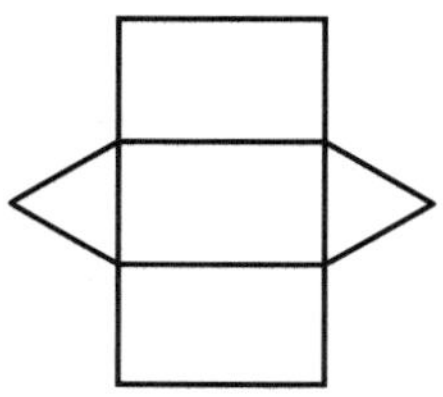	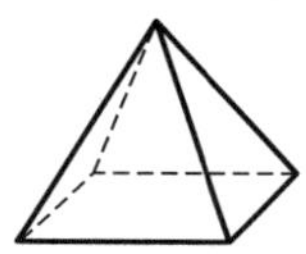 N	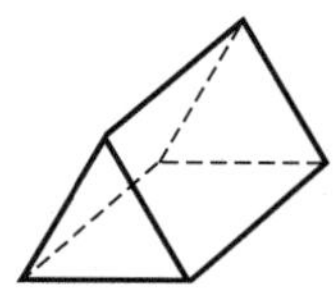 F	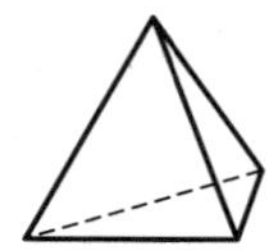 B
4. 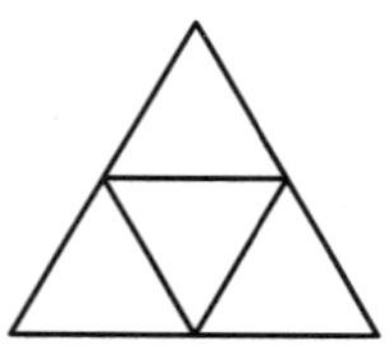	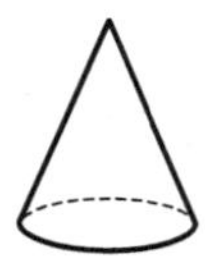 H	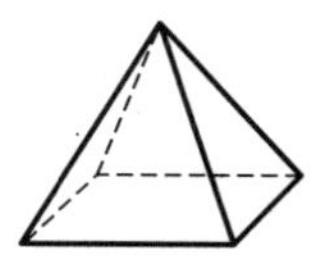 U	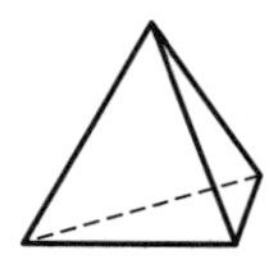 E
5. 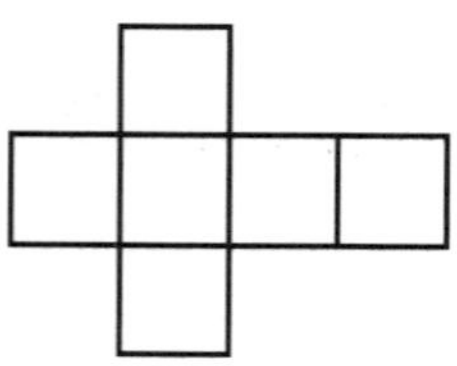	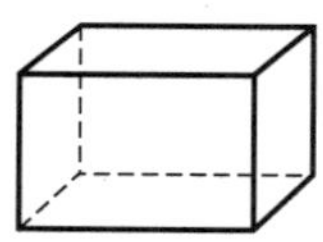 G	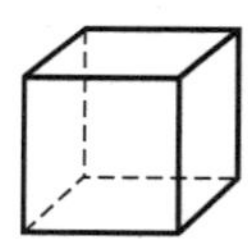 C	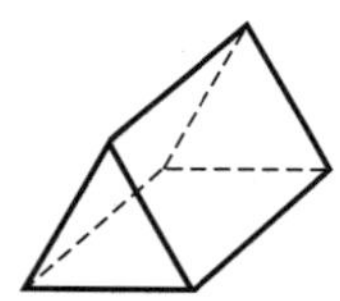 Y

Write the letter with a ring above the number of its row.

___ ___ ___ ___ ___
3 2 5 4 1

Name ______________________

Make It Right

♦♦♦

Identify the figure that matches the description. Find the box at the bottom of the page that has the name of the figure in it. Ring the word above the name of the figure.

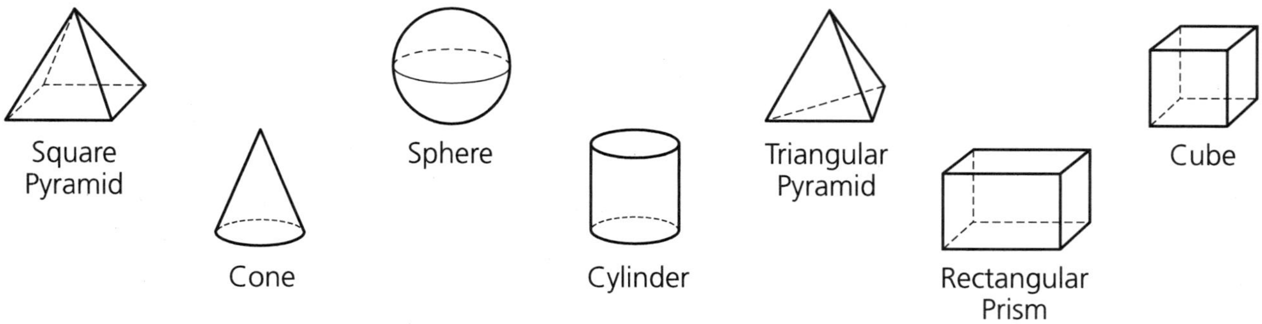

1. I have a circle on only one end. ______________________

2. I have 6 faces. All of them are squares. ______________________

3. I do not have any flat surfaces. ______________________

4. I have 5 faces. Four are triangular and one is square. ______________________

5. I have two circle faces, one at each end. ______________________

6. I have 6 faces. All of them are rectangles, and two of them are squares.

____

Which is correct: "9 and 5 *is* 13" or "9 and 5 *are* 13"?

BOTH hexagonal prism	NEITHER rectangular prism	NINE sphere	PLUS triangular pyramid
AND square pyramid	IS oval	FIVE cylinder	THE rhombus
ARE cube	THIRTEEN octagonal prism	CORRECT trapezoid	FOURTEEN cone

Write the words with a ring in order.

____

Name ______________________

Triangle Twins

Carefully cut out the nine triangles at the bottom of the page. Match each triangle cutout with a triangle at the top of the page that is exactly the same size and shape. Glue the cutout on top of the triangle it matches.

1. 2. 3. 4. 5.

6. 7. 8. 9.

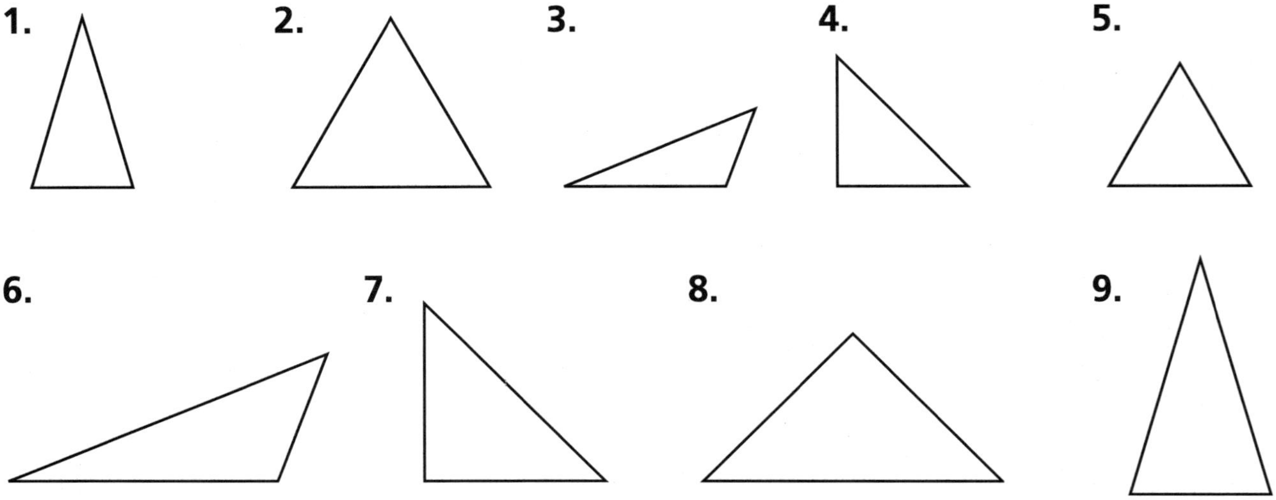

Write the letter found in each triangle in number order to spell out the word that tells how your pairs of triangles are related.

____

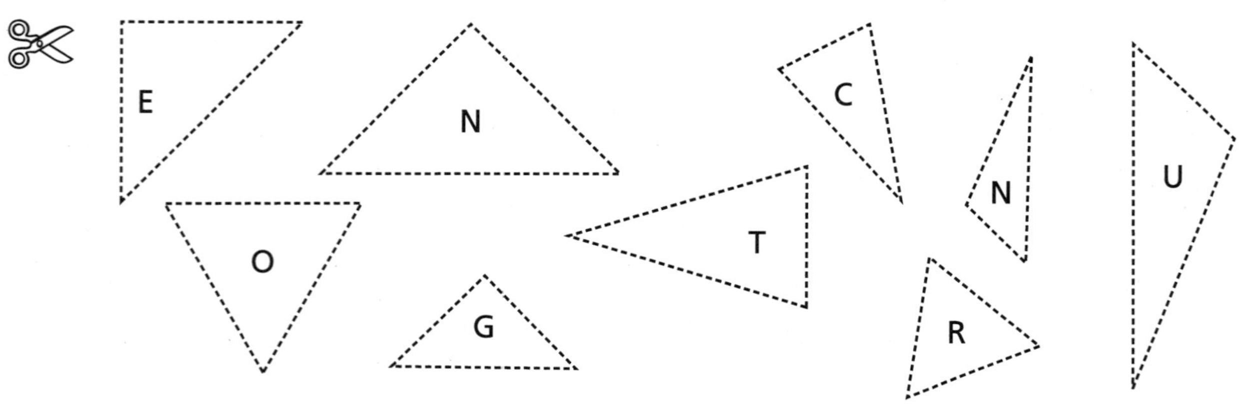

Name ______________________

Is There Symmetry?

Decide if each figure has a line of symmetry.
Ring the letter in the column that tells your answer.

	Figure	Yes	No		Figure	Yes	No
1.		T	W	2.		A	H
3.		E	S	4.	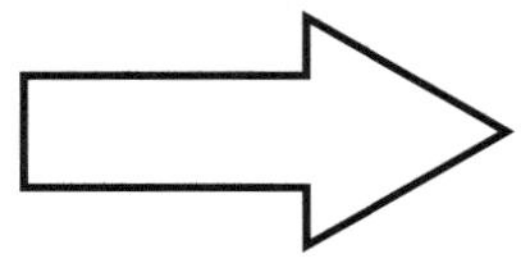	I	H
5.		T	A	6.		A	D
7.		T	B	8.	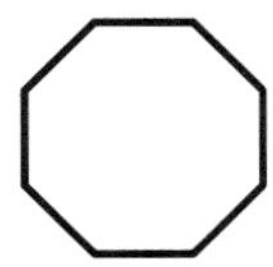	A	W
9.		O	T	10.	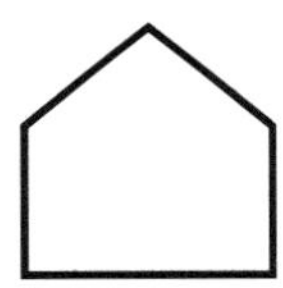	I	M
11.	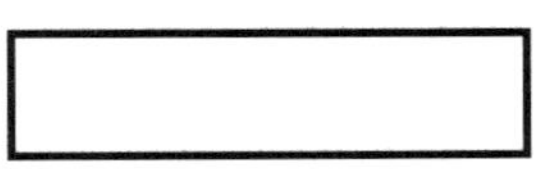	S	E	12.		H	A
13.		W	T				

Write the letters with a ring in order. The letters of your answer should read the same forward and backward.

____

Name ______________________

Reflect on Your Work

Shade squares in the box below each direction to show the filp.

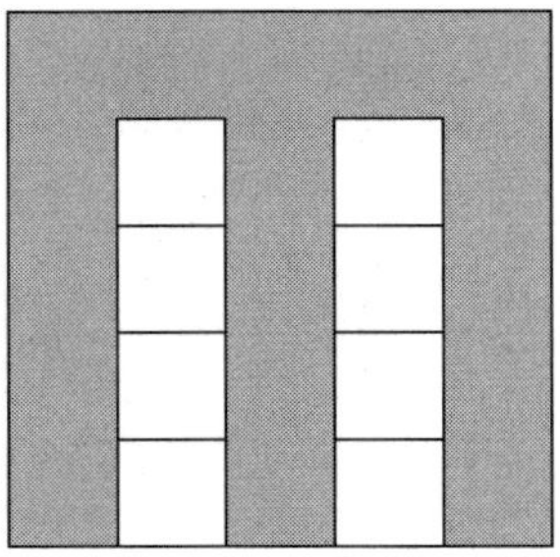

Flip top to bottom

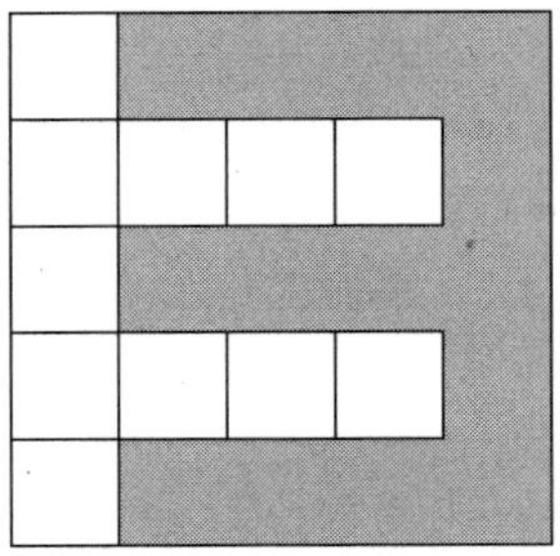

Flip left to right

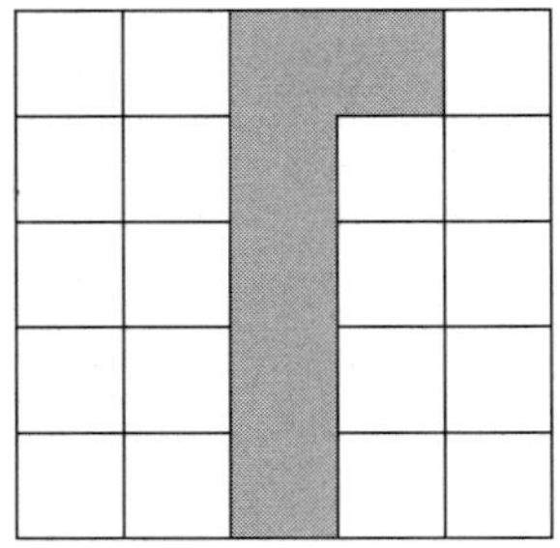

Flip top to bottom

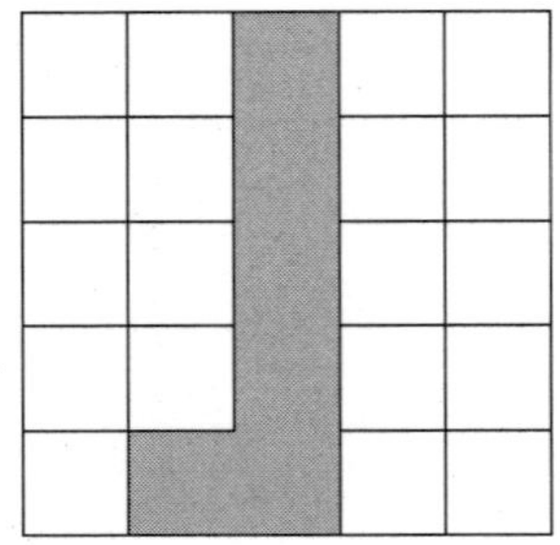

Flip left to right

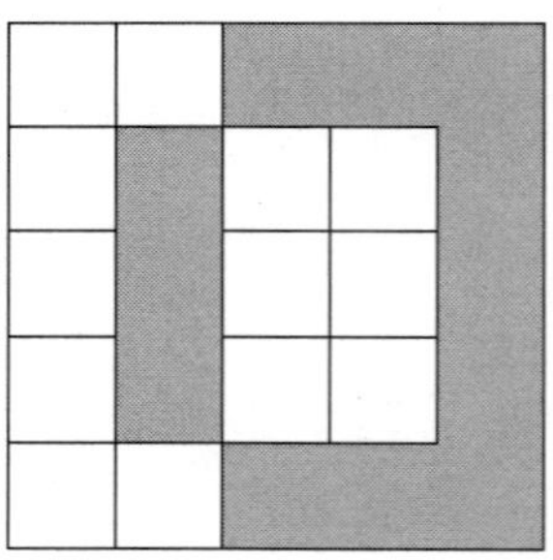

Flip left to right

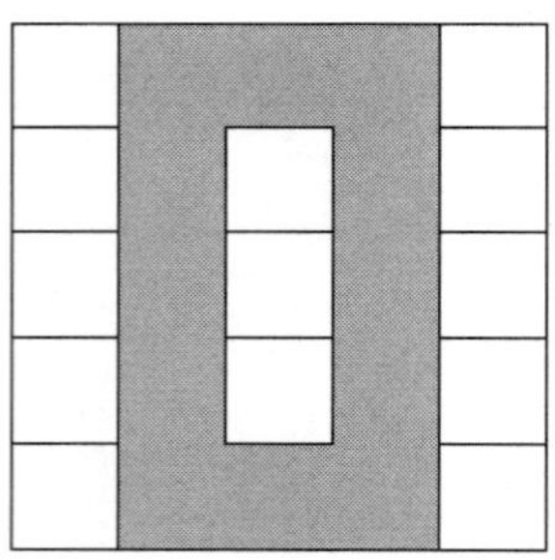

Flip top to bottom

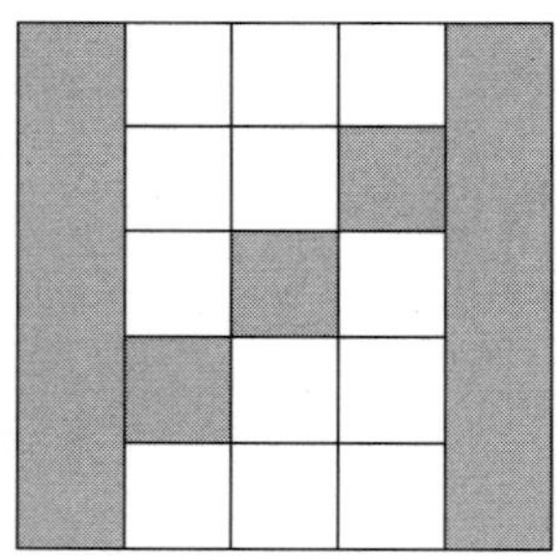

Flip left to right

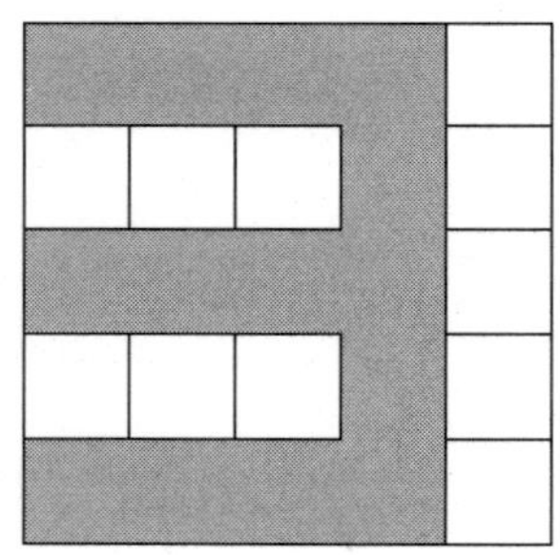

Flip left to right

Name ______________________________

How Do You Repair a Gorilla?

Write the letter of the point above its coordinates in the code.

___ (3,1) ___ (4,4) ___ (5,2) ___ (3,4)   ___ (1,5)   ___ (3,3) ___ (5,3) ___ (4,3) ___ (1,3) ___ (5,5) ___ (5,1)

___ (3,1) ___ (3,2) ___ (5,5) ___ (4,3) ___ (3,5) ___ (3,4)

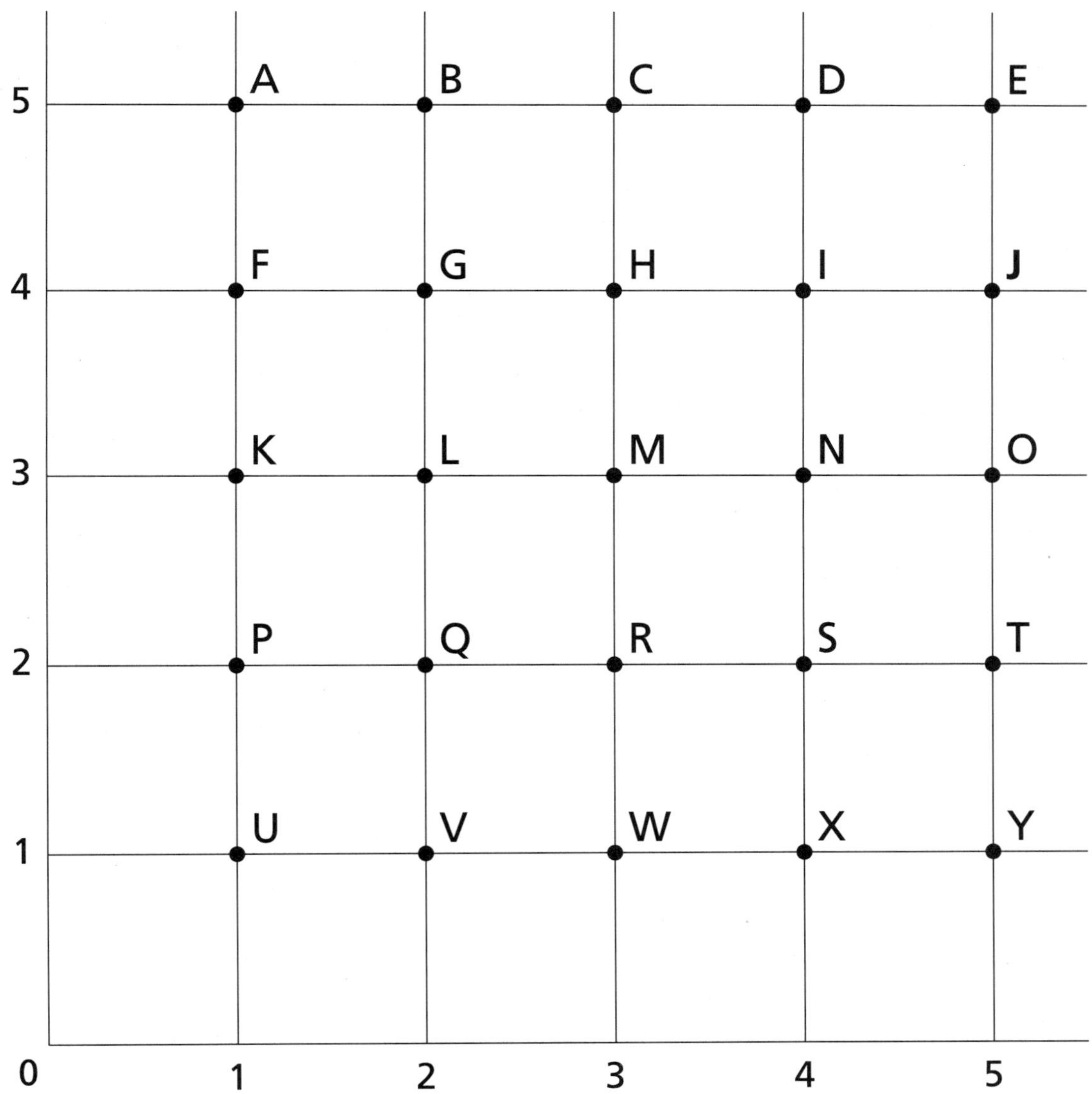

Name ______________________________

Sail Through Graphing

First graph each point on the coordinate grid.
Be sure to label it with its letter. Then connect the points in A,B,C order.

A (6,5) **B** (14,8) **C** (7,17) **D** (6,4) Lift your pencil

E (5,16) **F** (1,12) **G** (1,8) **H** (6,6) **I** (4,11) **J** (6,17) Lift your pencil

K (1,5) **L** (4,4) **M** (14,4) **N** (13,2) **O** (3,2) **P** (2,4)

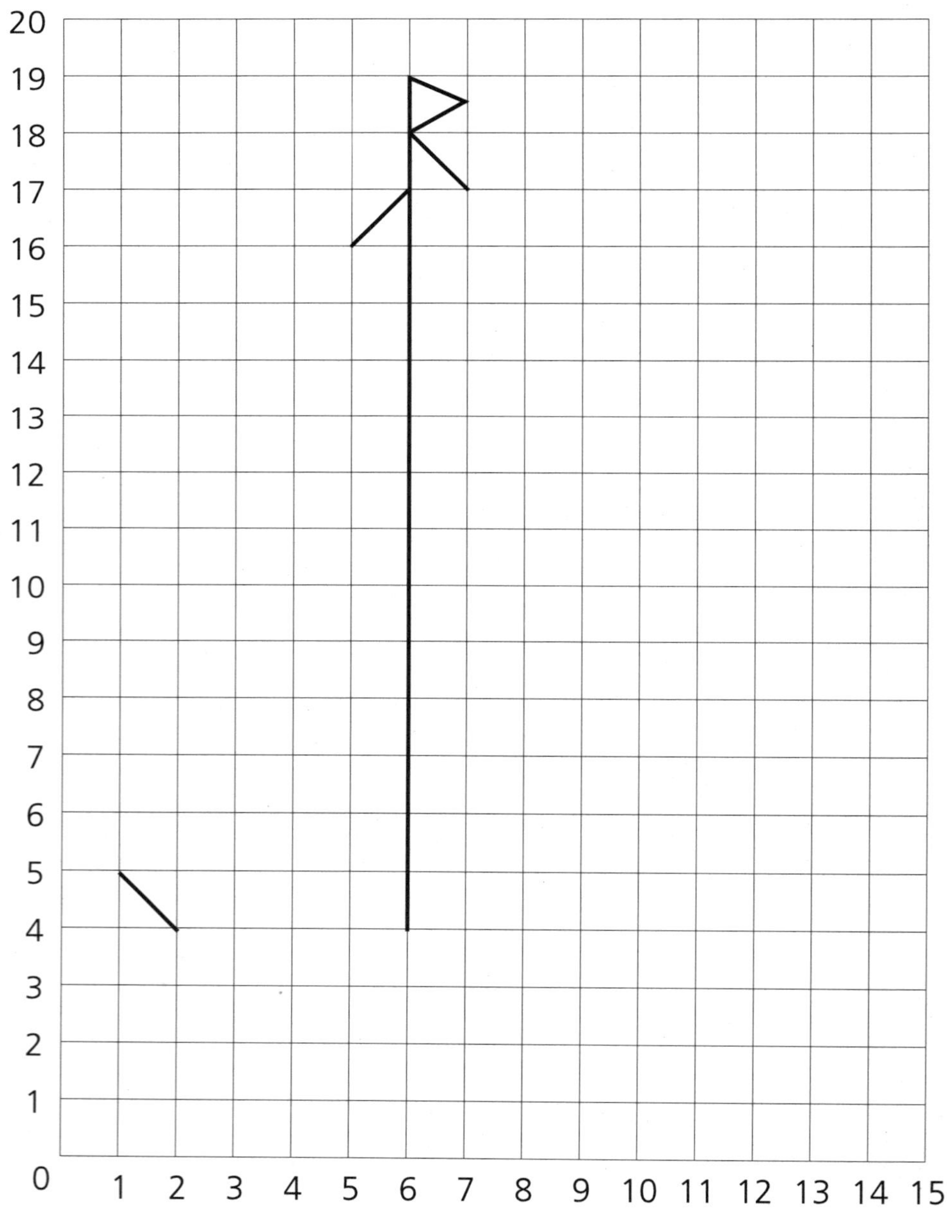

Name ______________________

Something to Crow About

Identify the rule of each pattern.
Then follow the directions in the table and color.

Rule	Color
Add 3 or subtract 3	Orange
Add 9 or subtract 9	Red
Multiply by 5 or divided by 5	Brown
Add 7 or subtract 7	Blue
Multiply by 2 or divided by 2	Green

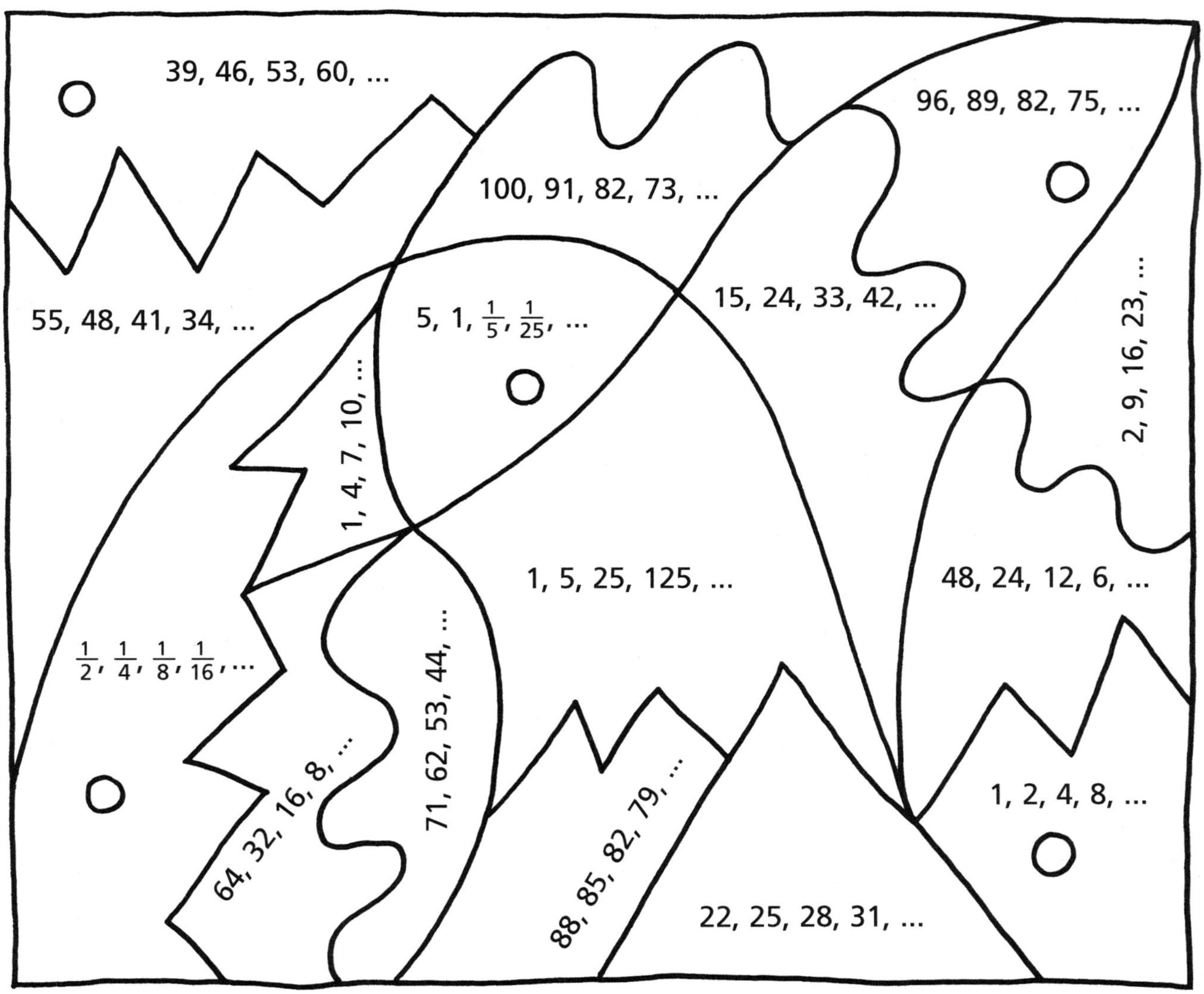

Name ______________________________

What Is Black and White and Red All Over?

A separate pattern is shown with each path of arrows.
Follow the arrows and fill in the missing numbers.

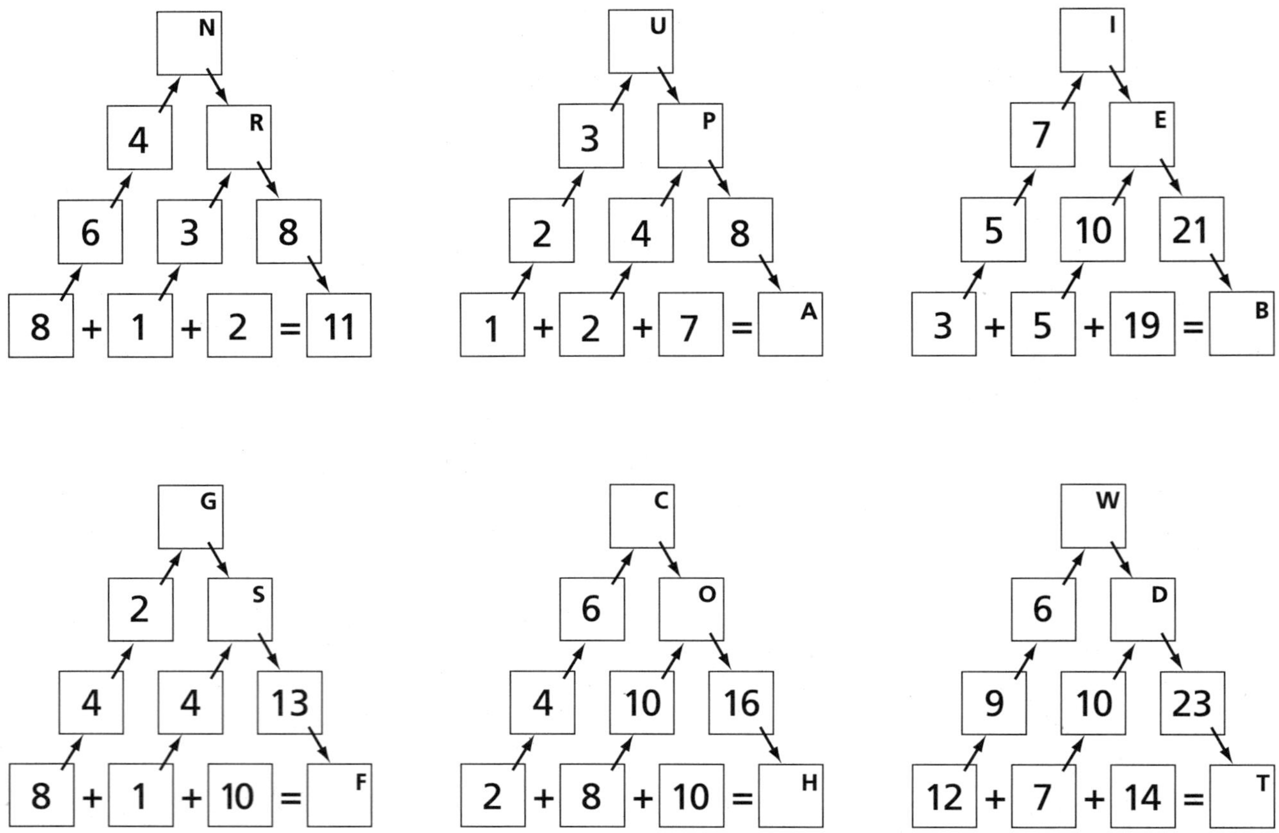

The numbers in the code are all numbers you have written in the pyramids. Each number has a letter in its box. Match the letter to the number in the code.

__ 10   __ 6 __ 15 __ 2 __ 1 __ 4 __ 9 __ 2   __ 3 __ 9 __ 33 __ 20   __ 10   __ 27 __ 10 __ 13

__ 8 __ 10 __ 7 __ 15   __ 12 __ 19   __ 7 __ 4 __ 2 __ 27 __ 4 __ 5 __ 2

Name ____________________

What Always Ends Everything?

Follow the directions for each box. Write the numbers from the box in order on the lines below the box. Then fill in the final number. Be careful, the directions are not always the same.

1. Count forward.

86		116
	106	
126		96

86 ____ ____ ____ ____ ____ G

2. Count backward.

1,998		2,002
	1,996	
2,000		1,994

____ ____ ____ ____ ____ ____ L

3. Count forward.

86		66
	46	
126		106

____ ____ ____ ____ ____ ____ T

4. Count backward.

1,287		1,387
	1,337	
1,237		1,187

____ ____ ____ ____ ____ ____ E

5. Count forward.

437		837
	237	
637		37

____ ____ ____ ____ ____ ____ R

6. Count backward.

5		4
	$5\frac{1}{2}$	
$3\frac{1}{2}$		$4\frac{1}{2}$

____ ____ ____ ____ ____ ____ H

Each answer has a letter below it.
Match the letter with the answer in the code.

___	___	___		___	___	___	___	___	___		___
146	3	1,137		1,992	1,137	146	146	1,137	1,037		136

Name ____________________

Growing with Patterns

Note: Use with page 61.

Draw the 4th figure in the pattern. Count how many squares you drew. Write that number in the table.

1.

Figure 1 Figure 2 Figure 3 Figure 4

Figure	Number of Squares
1	1
2	2
3	3
4	4 → **F**

2.

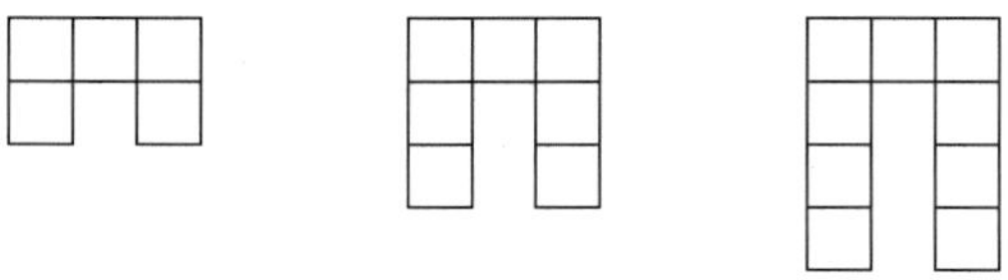

Figure 1 Figure 2 Figure 3 Figure 4

Figure	Number of Squares
1	5
2	7
3	9
4	______ → **P**

3.

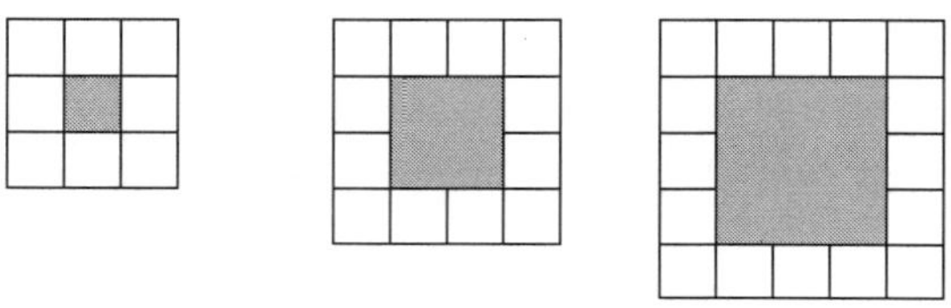

Figure 1 Figure 2 Figure 3 Figure 4

Figure	Number of Squares
1	8
2	12
3	16
4	______ → **S**

4.

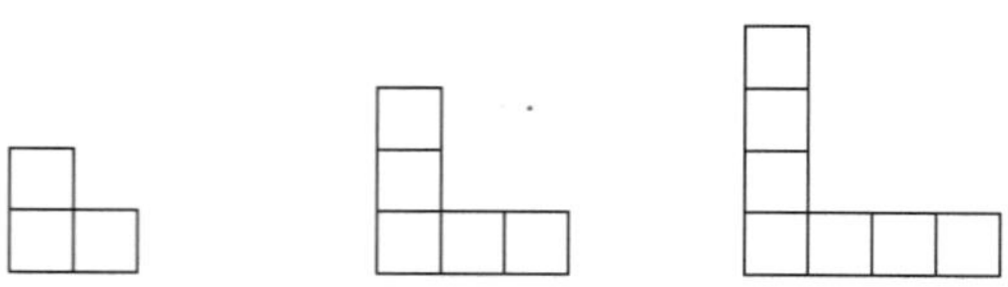

Figure 1 Figure 2 Figure 3 Figure 4

Figure	Number of Squares
1	3
2	5
3	7
4	______ → **N**

5.

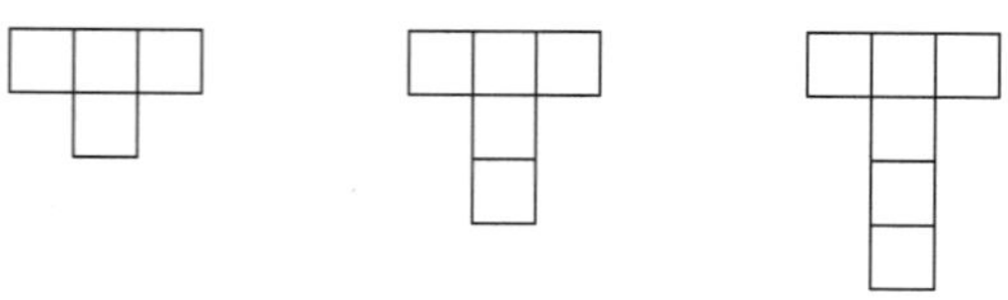

Figure 1 Figure 2 Figure 3 Figure 4

Figure	Number of Squares
1	4
2	5
3	6
4	______ → **Y**

6.

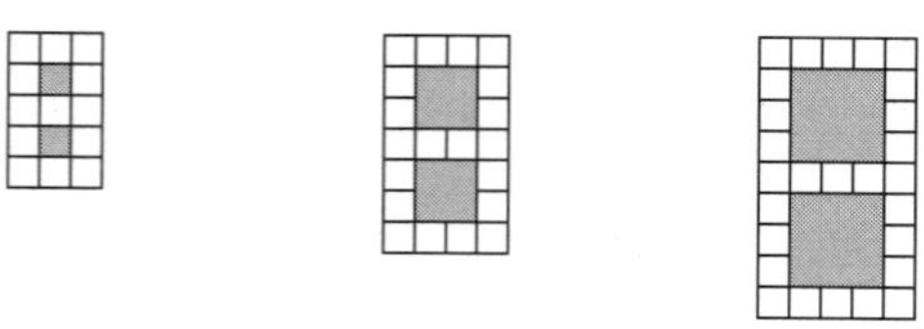

Figure 1 Figure 2 Figure 3 Figure 4

Figure	Number of Squares
1	13
2	20
3	27
4	______ → **R**

► Extend geometric growing patterns

Name ______________________

Growing with Patterns

Note: Use with page 60.

7.

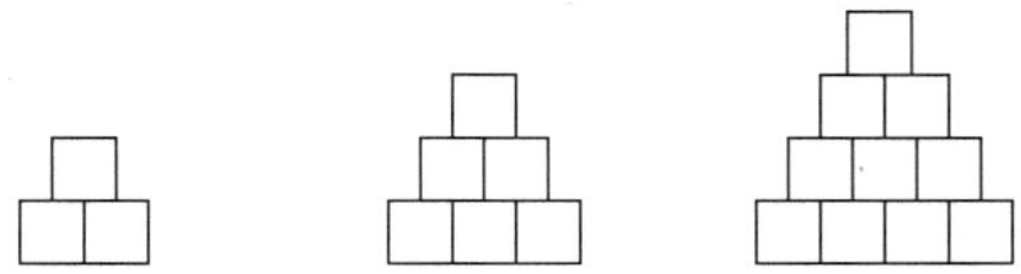

Figure 1 Figure 2 Figure 3 Figure 4

Figure	Number of Squares
1	3
2	6
3	10
4	______ → **I**

8.

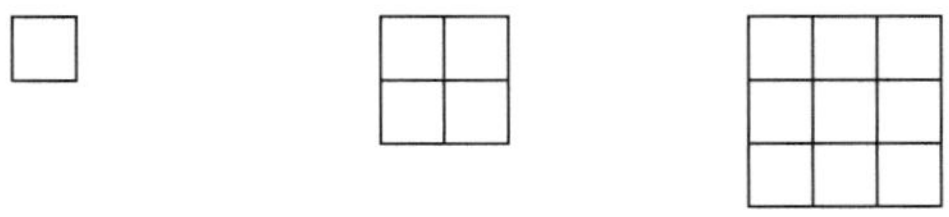

Figure 1 Figure 2 Figure 3 Figure 4

Figure	Number of Squares
1	1
2	4
3	9
4	______ → **O**

9.

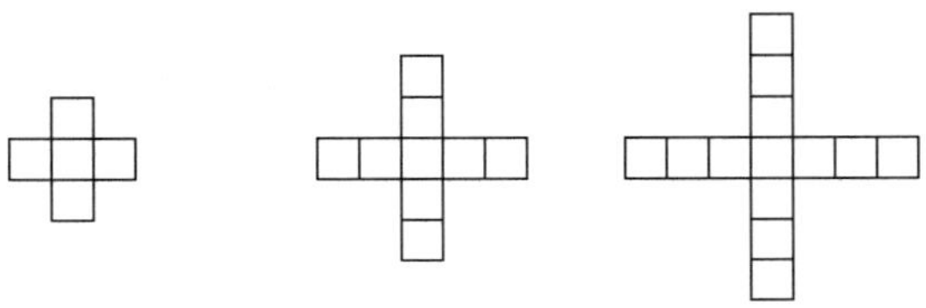

Figure 1 Figure 2 Figure 3 Figure 4

Figure	Number of Squares
1	5
2	9
3	13
4	______ → **T**

10.

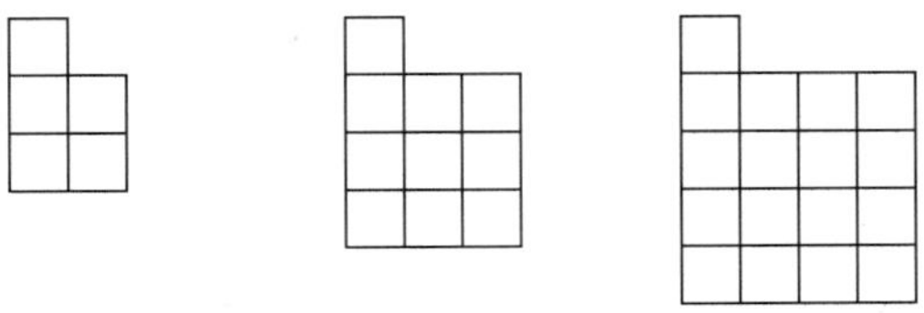

Figure 1 Figure 2 Figure 3 Figure 4

Figure	Number of Squares
1	5
2	10
3	17
4	______ → **A**

What has 100 legs but cannot walk?

Every number you wrote in a table has a letter after it. Match the letter with its number in the code.

___	___	___	___	___		___	___	___	___	___
4	15	4	17	7		11	26	15	34	20

___	___		___	___	___	___	___
16	4		11	26	9	17	20

Name ______________________

The Missing Link

In each sentence a shape takes the place of a number.
Find the value of that shape.

1. $3 + \square = 12$

$\square =$ ______ **E**

2. $\bigcirc \div 5 = 2$

$\bigcirc =$ ______ **E**

3. $\triangle + \triangle + 6 = 14$

$\triangle =$ ______ **N**

4. $\square + \square + \square + 9 = 15$

$\square =$ ______ **L**

5. $\bigcirc \div 2 = 10$

$\bigcirc =$ ______ **V**

6. $\triangle + 6 = 8 + 3$

$\triangle =$ ______ **O**

7. $9 \times \square = 54$

$\square =$ ______ **E**

8. $\bigcirc + \bigcirc - 3 = 11$

$\bigcirc =$ ______ **P**

What starts with "E" and ends with "E", but only has one letter?

Find each answer in the code. Write its letter above it.

___	___	___	___	___	___	___	___
6	4	20	9	2	5	7	10

Name ______________________________

What Is Always in Front of You but Cannot Be Seen?

Ring the letter of the sentence that will correctly solve the problem. Then solve the problem. Find your answer in the code at the bottom of the page and write the letter with a ring above it.

1. Felipé ordered 36 peanut butter cookies and 24 chocolate chip cookies for the class party. How many cookies did he order in all?

   M 36 − 24 = ☐ **T** 36 + 24 = ☐ **Y** 36 ÷ 2 = ☐

2. Kendra brought 4 packs of cups. Each pack had 12 cups. How many cups did she bring?

   R 12 − 4 = ☐ **N** 12 + 4 = ☐ **H** 4 × 12 = ☐

3. Kiona needed to collect $30 for the party. So far, she has collected $26. How much more does she need to collect?

   L 30 + 26 = ☐ **P** 30 ÷ 26 = ☐ **R** 30 − 26 = ☐

4. Eight children shared 2 pizzas. Each pizza had 12 slices. How many slices did each child get?

   U 24 ÷ 8 = ☐ **D** 12 × 8 = ☐ **M** 12 ÷ 8 = ☐

5. The desks were arranged in groups of 4. There were 7 groups of desks. Someone sat at each desk. How many children were in the class?

   R 4 + 7 = ☐ **F** 4 × 7 = ☐ **A** 7 − 4 = ☐

6. Telka made 18 party favors, and Caleb made 5 less than that. How many favors did Caleb make?

   E 18 − 5 = ☐ **I** 18 ÷ 5 = ☐ **W** 18 + 5 = ☐

___ ___ ___   ___ ___ ___ ___ ___ ___
60 48 13   28 3 60 3 4 13

Name ______________________

Missing Number Dot-to-Dot

Find the missing number in each sentence.

A	$12 + \square = 32$	**B**	$\square \div 4 = 6$
C	$11 \times \square = 66$	**D**	$23 - \square = 13$
E	$9 + \square = 31$	**F**	$42 \div \square = 6$
G	$\square \times 9 = 72$	**H**	$41 - 11 = \square$
I	$\square + 112 = 130$	**J**	$\square - 100 = 200$
K	$2 \times 7 + \square = 17$	**L**	$\square - 1 + 8 = 28$
M	$3 \times \square + 9 = 36$	**N**	$\square \div 10 = 9$
O	$\square - 40 = 60$	**P**	$\square + 59 = 100$

Write the letter of the sentence above its answer in the Code Box.

Code Box

300	90	22	24	9	18	100	21	10	7	30	41	20	3	8	6

Complete the dot-to-dot.
Follow the order of the letters
in the Code Box.

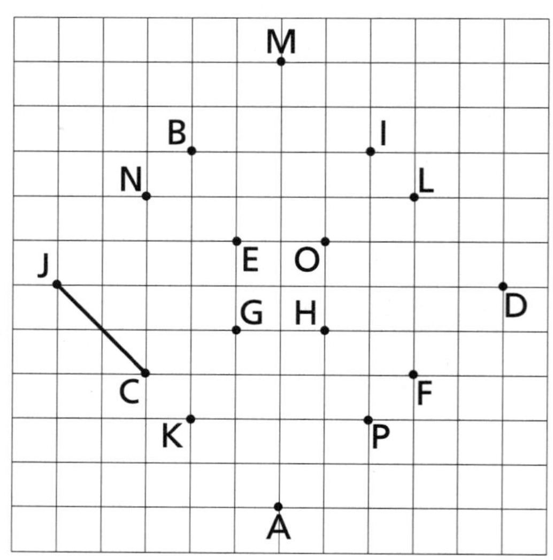

Name ____________________

What Could It Be?

Note: Use with page 66.

Write the missing numbers in each table.
Write the rule.

1.

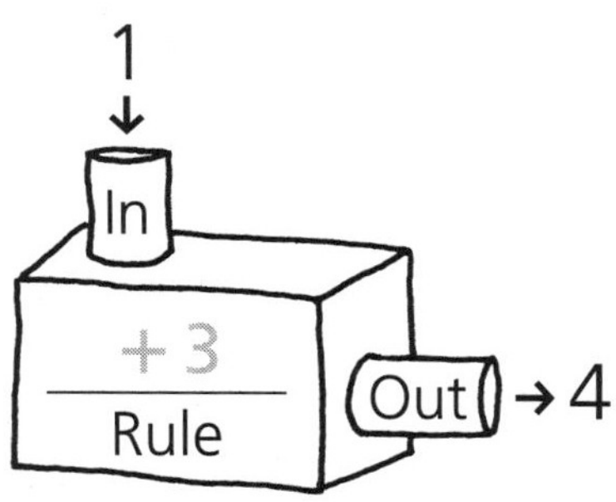

In	Out
3	6
5	8
7	10
9	
11	14
13	

2.

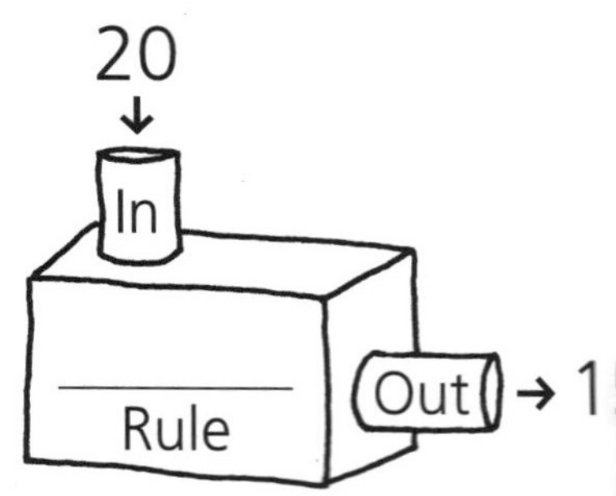

In	Out
100	95
90	85
	75
70	
60	55
50	45

3.

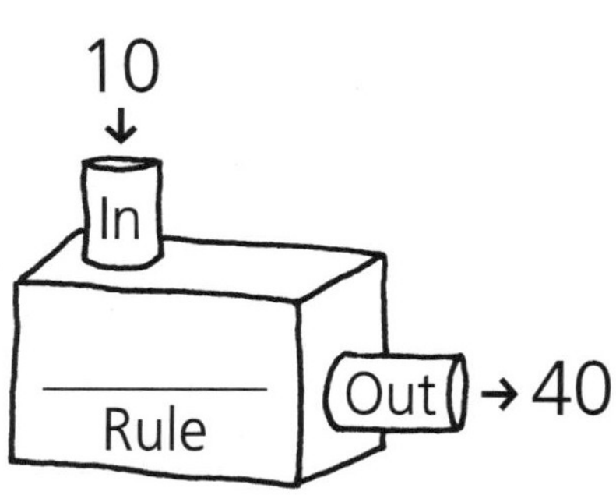

In	Out
2	8
3	12
4	16
5	
6	
7	

4.

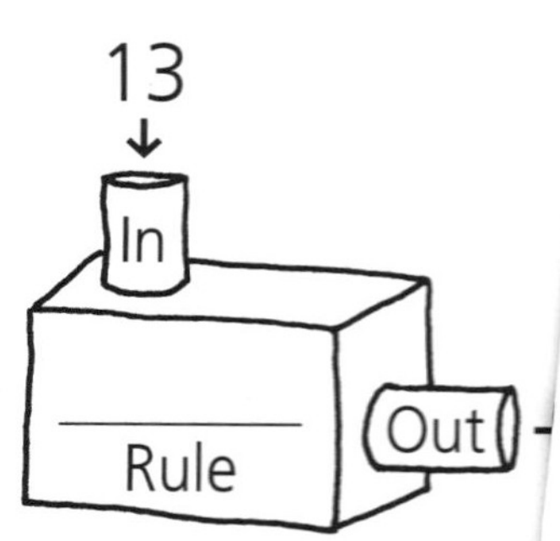

In	Out
35	41
39	
43	49
	53
51	
55	61

Name ______________________________

What Could It Be?

Note: Use with page 65.

5

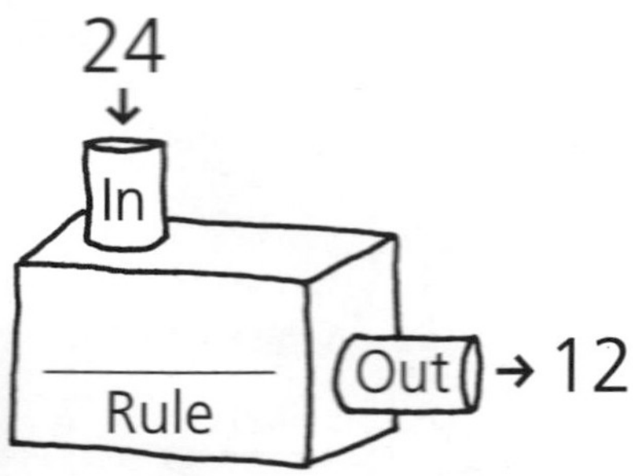

In	Out
12	6
22	11
32	16
42	
	26

6.

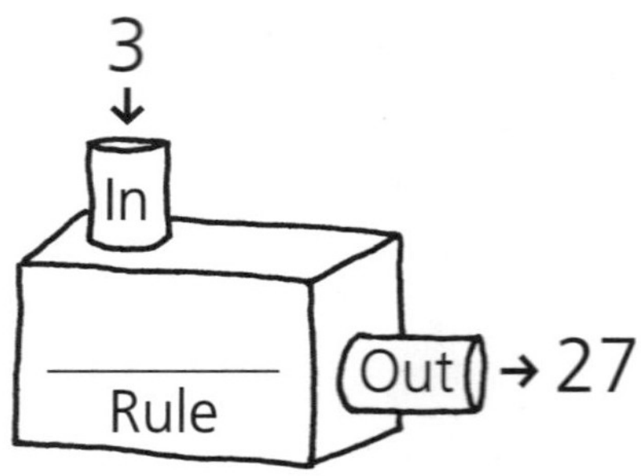

In	Out
2	18
4	36
6	
8	
10	90

as bright orange and sounds like a parrot?

each section that has a number you wrote in one of the tables.

Name ____________________

What Can You Always Count On?

Answer the questions below each graph.
Then place the letter of each exercise above its answer in the code.

Chores at Home

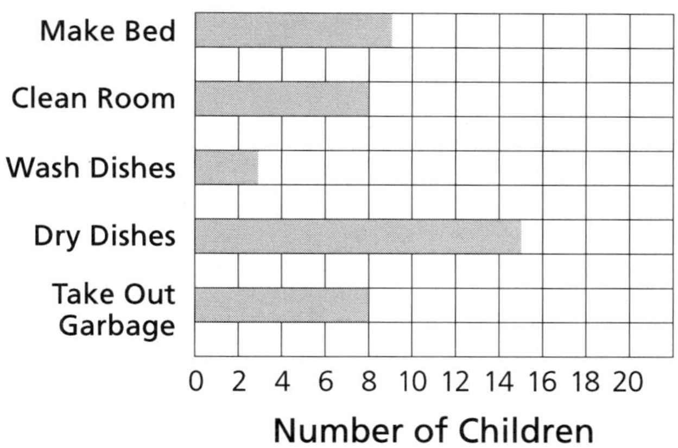

R How many children clean their room? ________

F How many children either make their bed or clean their room?

____

U If each child in this survey only named one chore, how many children participated in the survey?

____

S How many more children make their beds than wash dishes?

____

Y How many fewer children wash dishes than dry dishes? ________

Pepper Plant Growth

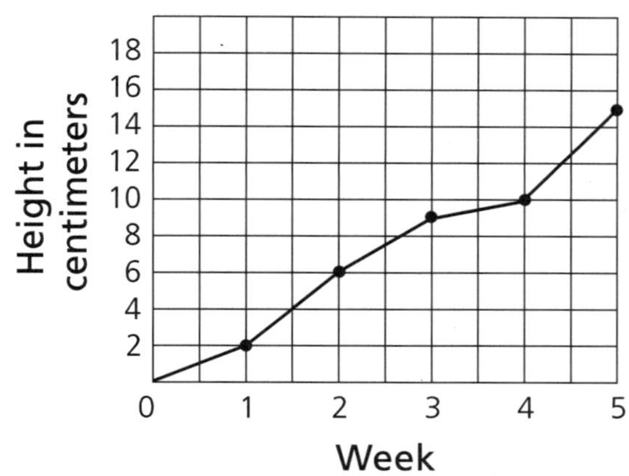

E How many centimeters did the pepper plant grow from week 1 to week 2? ________ cm

O What was the height of the pepper plant on week 4? ________ cm

I How many centimeters did the pepper plant grow from week 4 to week 5? ________ cm

G What was the least amount of growth during a one-week period for the pepper plant? ________ cm

N From week 1 to week 5, how many centimeters did the pepper plant grow? ________ cm

___	___	___	___		___	___	___	___	___	___	___
12	10	43	8		17	5	13	1	4	8	6

Name ______________________

A "Most Likely" Maze

Suppose you either spin a spinner or pick a marble from a bag without looking. Follow the path of the result that would be most likely.

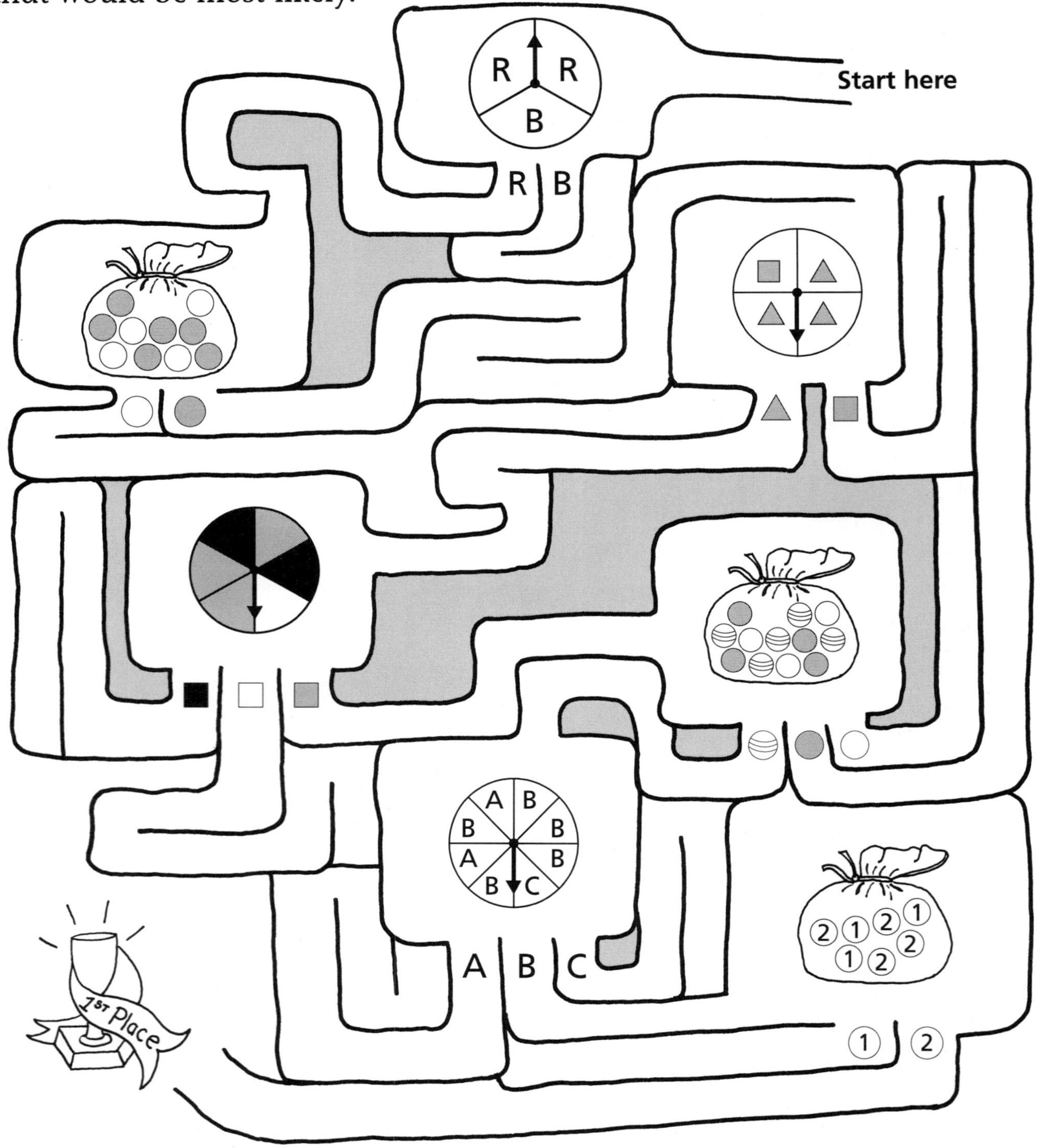

Name ______________________________

What Do You Think?

Ring the letter in front of the correct answer.

1. △ ○ □ △ ○ □

The shape that comes next is

K △ **M** ○

2. $\triangle \div 2 = 12$

A △ is greater than 10.

R △ is less than 10.

3. 51, 53, 55, 57...

L The next number will be even.

T The next number will be odd.

4. $5 + \square > 10$

O □ can be 5.

E □ can be greater than 100.

5. 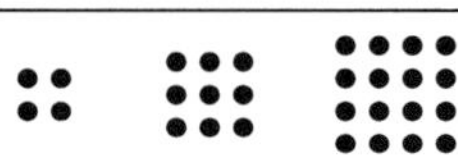

The next figure will have

D more than 20 dots

Y fewer than 20 dots

6. $\triangle \times 9 = 108$

G $\triangle < 10$

I $\triangle > 10$

7. odd number + even number = △

K △ is an even number.

U △ is an odd number.

8. even number + □ = even number

P □ is an even number.

N □ is an odd number.

9. 

The bottom number is

S odd **C** even

10. 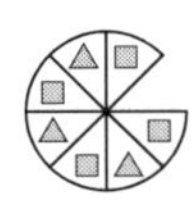

The shape on the missing piece is

Q ▲ **D** ●

Did you hear about the duck that flew upside down?

Match the letter with a ring to the number of its exercise in the code.

___ ___ ___ ___ ___ ___ ___ ___ ___ ___ ___.
6 3 10 7 2 9 1 4 5 7 8

Name ______________________________

What Has Many Ears but Cannot Hear?

Use a ruler. Draw a straight line from each event to **more likely**, **less likely**, or **equally likely** to show the chances that the event will happen. Your lines will cross out letters. Write the letters that remain in order at the bottom of the page.

1. Omar rolls a number cube. He thinks the number that lands on top will be greater than 2. →
2. Zena flips a coin. She thinks it will land heads up. →
3. Nadia has a bag with 20 red cubes and 10 blue cubes. She thinks if she picks a cube without looking it will be red. →
4. The school bus is late about once a week. Quan thinks the bus will be late on Monday. →
5. There are 1 blueberry and 2 carrot muffins in a bag. Erika takes a muffin from the bag without looking. She thinks it will be a carrot muffin. →
6. The names of 12 boys and 12 girls are placed in a box. Doug picks a name from the box without looking. He thinks he will pick a girl's name. →
7. Peter packs a pickle with his lunch every day. He will pack a pickle with his lunch tomorrow. →
8. Duane has 10 cards numbered from 1 to 10. Jody picks a card without looking. She thinks she will pick a 7. →

B

S T O ← **More Likely**

R A L

O M K

S

P I E ← **Less Likely**

O R F

J M

C O T

J

E H Y ← **Equally Likely**

A R

N

____

Name ______________________________

Think of What Is Possible

Follow the directions in each box. Find all possible answers.

1. **2, 3, 4** Multiply 2 of the numbers. Then add the third number to the product.	**2.** **1, 5, 9** Make 3-digit numbers. Use all 3 digits in each number you make.
3. **2, 4, 6, 3, 0** Add only 2 of the numbers. Find the largest sum. Find the smallest sum.	**4.** **4, 6, 8** Make the 3 largest numbers possible. Use all 3 in each number you make.
5. **2, 3, 7** Multiply 2 of the numbers. Then add the third number to the product.	**6.** **2, 4, 6** Add 2 of the numbers. Then multiply the sum by the third number.
7. **1, 2, 3, 4** Make the 3 smallest numbers possible. Use all 4 digits in each number you make.	

How can you tell if an elephant is getting ready to charge?

Possible answers for each exercise are grouped together. Ring only those that are correct. Then write the letters above the numbers with a ring in order.

S	H	R	E	T
34	36	24	32	20

H	A	K	E	S
3,124	1,234	1,324	1,243	1,342

S	O	D	U
10	11	12	14

T	H	I
864	684	846

S	C	R	E	D	I
159	519	915	951	591	195

T	G	C	A
13	20	23	17

L	R	D
9	10	2

____.

Name ____________________

What Are the Chances?

Decide the probability of each event that is described.
Write the probability as a fraction.

N You roll the cube. It will land with 5 on top.

______

H You spin the spinner. It will land on Y

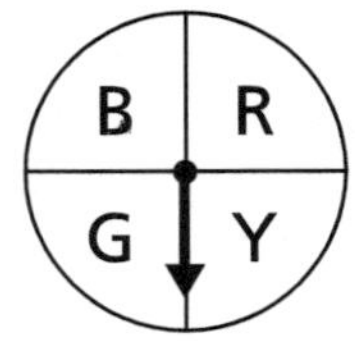

______

R You pick a marble without looking. It will be a black marble.

______

E You flip a coin. It will land heads up.

______

D You spin the spinner. It will land on △.

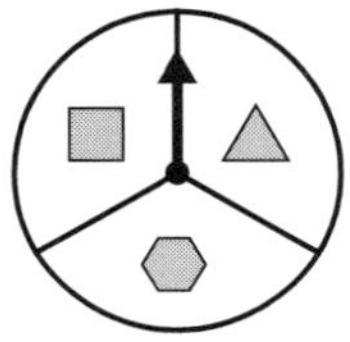

______

G You pick a tile without looking. It will **not** be a white tile.

______

C You roll the cube. It will land with a number that is **not** 6 on top.

______

S You pick a crayon without looking. It will be green.

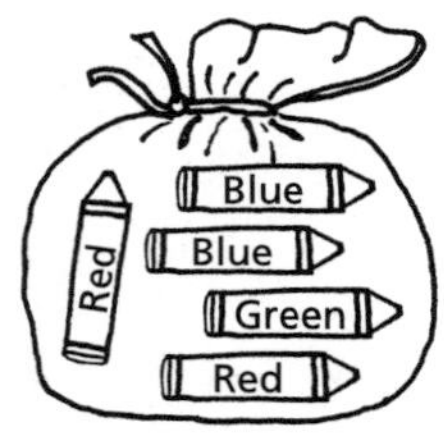

______

O You take a piece of fruit without looking. It will be an apple.

______

A You roll the cube. It will land with a number that is less than 7 on top.

______

Match the letter of the exercise with its answer in the code.

___ ___ ___ ___ ___ ___ ___   ___ ___ ___   ___ ___ ___ ___ .

5/6 1/4 6/6 1/6 5/6 1/2 1/5   6/6 3/5 1/2   2/3 3/4 3/4 1/3

► Read and write numbers Name

Blast Off!

Follow each direction as you advance through the maze.

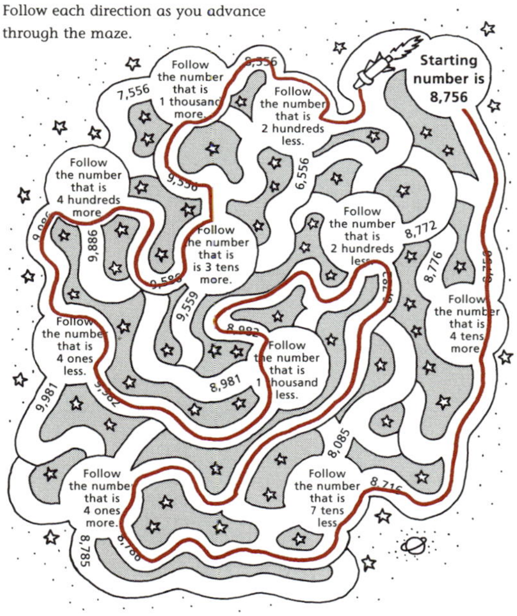

► Read and write numbers Name

How Is Your Aim?

Use a ruler to draw a line from the tip of each arrow to the center of the target with the same number. Your lines will cross out letters. Write the letters that are left in order at the bottom of the page.

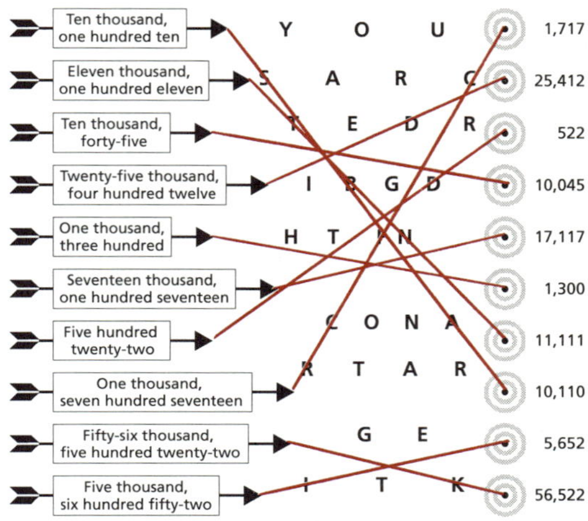

YOU ARE RIGHT ON TARGET

(YOU ARE RIGHT ON TARGET.)

► Read and write numbers Name

The "Write" Numbers

Write a numeral for each number word.
Then color the section of the design that matches your numeral.

1. Seventeen thousand, one **17,001**
2. Forty-five thousand, seven hundred forty-one **45,741**
3. Twenty-six thousand, four hundred fifty-three **26,453**
4. Ninety-five thousand, five **95,005**
5. Sixty thousand, one hundred three **60,103**
6. Nine thousand, ten **9,010**
7. Twenty-one thousand, six hundred fifty-four **21,654**
8. Nineteen thousand, thirty **19,030**
9. Six thousand, fifty-two **6,052**
10. Twenty-five thousand, one hundred three **25,103**
11. Sixty thousand, five hundred fifty-five **60,555**
12. Forty thousand, two hundred thirty-four **40,234**

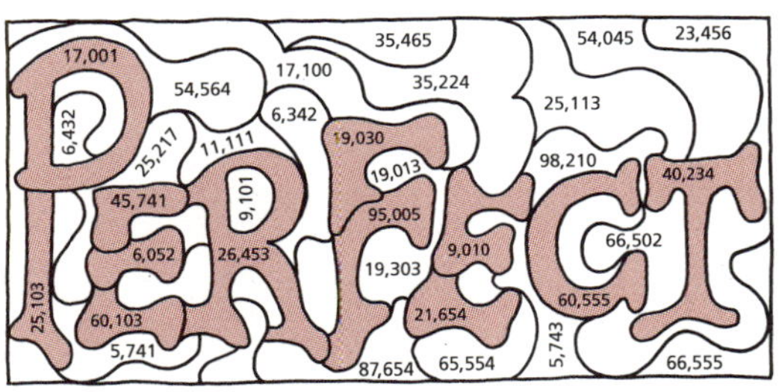

► Read and write numbers Name

How Hard Was It to Decode This Message?

Write the number.

	A	E	I	O	B	T
176,142 + 5 hundreds	1	7	6	6	4	2
Eight hundred twelve thousand, three hundred	8	1	2	3	0	0
Eighteen thousand, two hundred fifty-four		1	8	2	5	4
Nine thousand, three hundred twenty-three			9	3	2	3
3 hundreds + 8 tens + 3 ones				3	8	3

	H	C	W	Y	D	S
623,211 + 6 ten thousands	6	8	3	2	1	1
One hundred fifty-one thousand, one hundred	1	5	1	1	0	0
Thirty-five thousand, seven hundred seventeen		3	5	7	1	7
Eighty thousand, six hundred seventy-eight		8	0	6	7	8
632 + 4 tens				6	7	2

Find new numbers by reading the columns.
Match the letter of the column to the number in the code.

T (20,433) H (61) I (6,289) S (10,782) W (3,150) A (18) S (10,782) A (18) S (10,782)

E (711) A (18) S (10,782) Y (21,766) A (18) S (10,782) A (18) - B (40,528) - C (8,538)

T (20,433) O (63,233) D (10,177) E (711) C (8,538) O (63,233) D (10,177) E (711).

► Compare whole numbers   Name ____________

Mole Hole

Follow the directions in each circle to help the mole find his home.

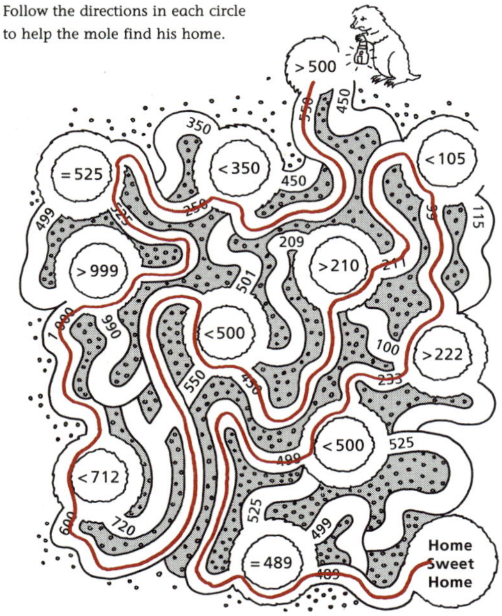

► Compare whole numbers   Name ____________

An Evening Song

Fill in the ◯ to compare each pair of numbers.
Use < for less than, > for greater than, or = for equal.

Color spaces with > black.
Color spaces with < brown.
Color spaces with = yellow.

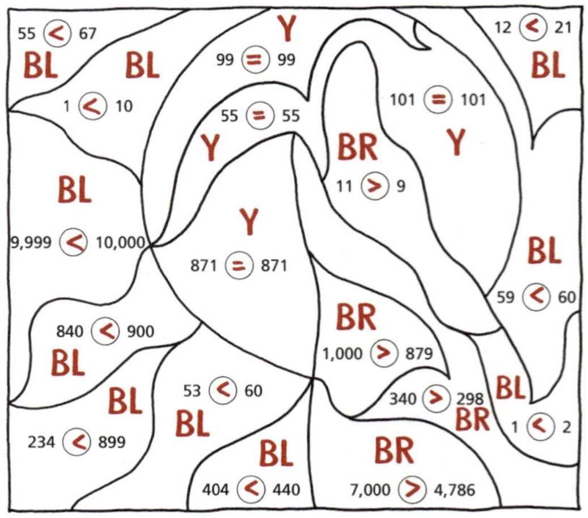

► Order whole numbers   Name ____________

Animal Abilities

Write the missing whole number. Then find the number in the code and write the letter above it.

1. I'm the number just before 100. 99 (N)
2. You'll find me between 999 and 1,001. 1,000 (H)
3. When you count, you say me right before 10,000. 9,999 (W)
4. I'm the 6th counting number you say after 595. 601 (T)
5. I'm the number between 817 and 819. 818 (O)
6. I come right before 889. 888 (E)
7. If you count back 2 numbers from 9,911, you'll find me. 9,909 (I)
8. I'm the number between 609 and 611. 610 (P)
9. You'll find me right before 910. 909 (G)
10. I'm the 3rd counting number you say after 805. 808 (S)

How does a skunk smell?

W	I	T	H		H	I	S		N	O	S	E
9,999	9,909	601	1,000		1,000	9,909	808		99	818	808	888

What do pigs use when they write letters?

P	I	G		P	E	N	S
610	9,909	909		610	888	99	808

► Round whole numbers   Name ____________

Leaping Lily Pads!

Round each number.

1. 156 to the tens place 160
2. 329 to the nearer hundred 300
3. 360 to the hundreds place 400
4. 72 to the nearer ten 70
5. 277 to the tens place 280
6. 609 to the hundreds place 600
7. 609 to the tens place 610
8. 691 to the tens place 690
9. 199 to the nearer hundred 200
10. 8 to the nearer ten 10
11. 18 to the tens place 20
12. 61 to the hundreds place 100

Complete the dot-to-dot. Follow your answers in order.

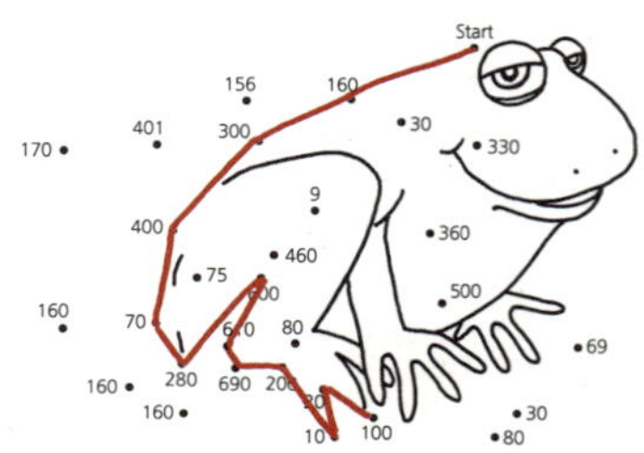

► Use front-end estimation

Name ____________

Estimation Puzzle

Note: Use with page 10.

Write the front-end estimate for each number.

14 → 10	829 → 800	56 → 50
72 → 70	32 → 30	156 → 100
604 → 600	94 → 90	211 → 200
64 → 60	433 → 400	85 → 80
26 → 20	362 → 300	
999 → 900	735 → 700	

Cut out these 16 squares to use on page 10.

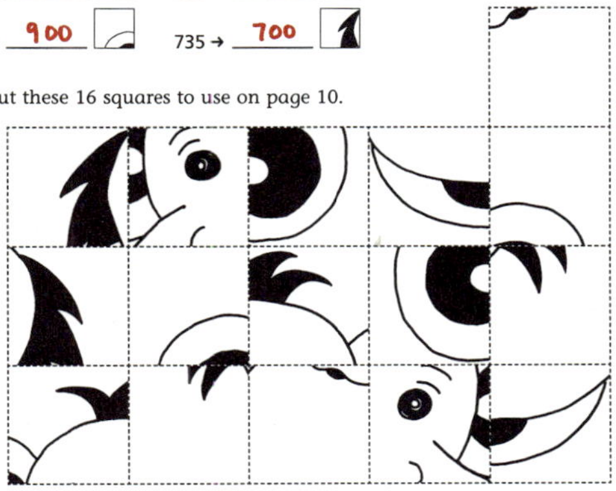

► Use front-end estimation

Name ____________

Estimation Puzzle

Note: Use with page 9.

Match each cutout square to one pictured on page 9.
Find the front-end estimate for that picture in the grid.
Paste the cutout in the square with the estimate.

► Practice addition and subtraction facts

Name ____________

Cross-Number Puzzles

Write a number in each box with a letter
to make true sentences that go across and down.

14	−	8	=	R 6
+	■	−	■	+
2	+	O 2	=	4
=	■	=	■	=
S 16	−	6	=	10

3	+	Z 7	=	10
+	■	+	■	+
4	+	V 1	=	5
=	■	=	■	=
7	+	U 8	=	15

13	+	6	=	I 19
+	■	−	■	+
A 3	−	2	=	1
=	■	=	■	=
16	+	4	=	H 20

D 18	−	6	=	12
−	■	+	■	−
4	+	8	=	Q 12
=	■	=	■	=
14	−	F 14	=	0

9	+	9	=	18
+	■	−	■	−
Y 5	+	T 4	=	N 9
=	■	=	■	=
14	−	5	=	9

2	+	8	=	E 10
+	■	−	■	+
L 13	−	6	=	7
=	■	=	■	=
15	+	2	=	P 17

Find each answer in the code. Write its letter above it.

Y	O	U		S	O	L	V	E	D
5	2	8		16	2	13	1	10	18

T	H	E	S	E		P	U	Z	Z	L	E	S
4	20	10	16	10		7	8	7	7	13	10	16

F	A	I	R		A	N	D		S	Q	U	A	R	E
14	3	19	6		3	9	18		16	12	8	3	6	10

► Practice multiplication and division facts

Name ____________

Dot-to-Dot

Solve each exercise. Then complete the dot-to-dot
by connecting your answers in order.

1. 18 ÷ 9 = 2
2. 3 × 6 = 18
3. 7 × 2 = 14
4. 20 ÷ 2 = 10
5. 16 ÷ 2 = 8
6. 18 ÷ 3 = 6
7. 4 × 5 = 20
8. 4 × 4 = 16
9. 3 × 4 = 12
10. 16 ÷ 4 = 4
11. 9 ÷ 3 = 3
12. 5 × 3 = 15
13. 3 × 3 = 9
14. 3 ÷ 3 = 1

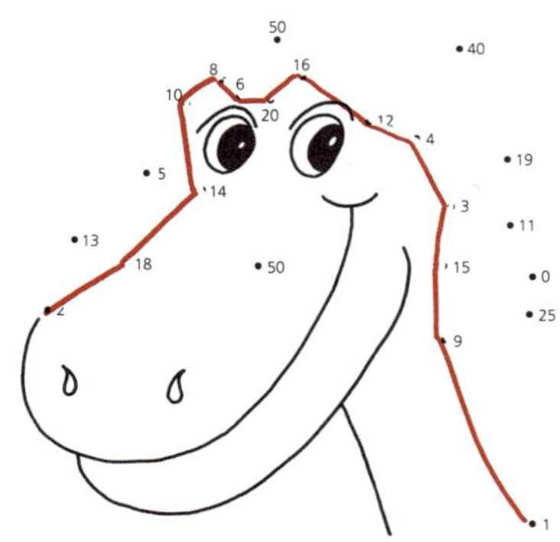

► Practice facts with mixed operations Name ____________

A Day at the Carnival

Solve each problem. Write the boxed word from the question above its answer in the code.

Ferris Wheel
5 tickets
per ride

1. Luis, Sophia, and Bihn purchased tickets [at] the carnival. Luis bought 6 tickets, Sophia bought 7, and Bihn bought 5. They decided to put their tickets together. How many tickets did the three friends have in all? **18** tickets
2. If [the] three friends shared the tickets equally, how many would each one get? **6** tickets
3. Luis held the tickets for everyone. He had a hole in his pocket and 4 tickets fell out. How many [were] left? **14** tickets
4. [When] they discovered that Luis had lost the 4 tickets, Sophia used the rest of her money to buy 2 more tickets. How many tickets did they have then? **16** tickets
5. The three friends decided to ride the Ferris wheel. How many tickets did [they] have to use? **15** tickets
6. How many tickets were left at the [end] of the day? **1** ticket

When did the three friends find the lines for the rides were the longest?

WHEN	THEY	WERE	AT	THE	END
16	15	14	18	6	1

► Practice facts with mixed operations Name ____________

Where Do Fish Keep Their Money?

Complete each number sentence.

E $20 \div 5 \times 2 =$ **8**

N $16 \div 4 \times 3 =$ **12**

V $4 \times 3 + 4 =$ **16**

K $13 - 3 \div 5 =$ **2**

S $6 + 8 - 5 =$ **9**

R $10 + 5 \div 3 =$ **5**

A $7 \times 2 + 5 =$ **19**

B $8 \times 3 \div 6 =$ **4**

I $18 - 7 \times 0 =$ **0**

Find each answer in the code.
Write the letter of the fish above it.

I	N		R	I	V	E	R		B	A	N	K	S
0	12		5	0	16	8	5		4	19	12	2	9

► Estimate sums Name ____________

What Can't You Eat for Lunch?

Estimate by rounding each number to the nearer ten.

E	62 + 38	(60) + (40) = **100**
O	29 + 52	(30) + (50) = **80**
S	39 + 53	(40) + (50) = **90**
A	96 + 28	(100) + (30) = **130**
E	42 + 57	(40) + (60) = **100**
R	21 + 93	(20) + (90) = **110**

Estimate by rounding each number to the nearer hundred.

K	422 + 683	(400) + (700) = **1,100**
U	927 + 463	(900) + (500) = **1,400**
B	701 + 159	(700) + (200) = **900**
T	481 + 252	(500) + (300) = **800**
F	395 + 239	(400) + (200) = **600**
P	719 + 322	(700) + (300) = **1,000**

Match the letter of the problem with its estimated answer in the code.

B	R	E	A	K	F	A	S	T
900	110	100	130	1,100	600	130	90	800

O	R		S	U	P	P	E	R
80	110		90	1,400	1,000	1,000	100	110

► Estimate differences Name ____________

Why Did the Leopard Wear a Mask?

Estimate each difference.
Then match the letter of the problem with its answer in the code.

N	33 − 14	30 − 10 = 20
H	72 − 39	70 − 40 = **30**
A	89 − 22	90 − 20 = **70**
E	53 − 14	50 − 10 = **40**
W	85 − 26	90 − 30 = **60**
D	94 − 77	90 − 80 = **10**
B	87 − 36	90 − 40 = **50**
P	593 − 246	600 − 200 = **400**
I	734 − 582	700 − 600 = **100**
T	365 − 249	400 − 200 = **200**
O	761 − 493	800 − 500 = **300**
S	816 − 297	800 − 300 = **500**

H	E		D	I	D		N	O	T
30	40		10	100	10		20	300	200

W	A	N	T		T	O		B	E
60	70	20	200		200	300		50	40

S	P	O	T	T	E	D
500	400	300	200	200	40	10

▶ Answers for pages 17–20

▶ Add and subtract without regrouping

Name ________________

Why Did Julio Collect Clocks?

Solve each problem. Write the boxed word from the question above its answer in the code.

1. Rafael likes [to] collect stamps from all over the world. He has 375 stamps in his collection. If 215 stamps are from the United States, how many are from other countries? **160** stamps
2. Sachi decided to [save] seashells. During her vacation at the beach she found 156 shells. She left 54 of the shells at the beach because they were broken. How many shells did she keep? **102** shells
3. Tommy spends a lot of [time] collecting trading cards. He has 623 baseball cards, 111 basketball cards, and 62 football cards. How many trading cards does Tommy have? **796** cards
4. Millie [wanted] to add to her collection of 100 buttons. A friend gave her 32 buttons, and her Aunt Tillie gave her 12 more. How many buttons did Millie have in her collection then? **144** buttons
5. Ina collected 586 coins. She gave away 120 coins [because] they were doubles. How many coins were left in her collection? **466** coins
6. Kwan had 378 miniature cars in his collection. [He] gave 130 away to his best friends Jamal and Henry. How many cars were left in Kwan's collection? **248** cars

BECAUSE	HE	WANTED	TO	SAVE	TIME
466	248	144	160	102	796

Elementary Math with *Pizzazz!* 17

▶ Add and subtract with regrouping

Name ________________

Hidden Picture

Find each sum or difference. Then color each area of the picture below that has one of your answers.

9 + 4	33 + 18	60 − 25	15 − 6	27 + 5
13	**51**	**35**	**9**	**32**

105 + 5	322 − 8	631 − 24	406 + 127	555 − 216
110	**314**	**607**	**533**	**339**

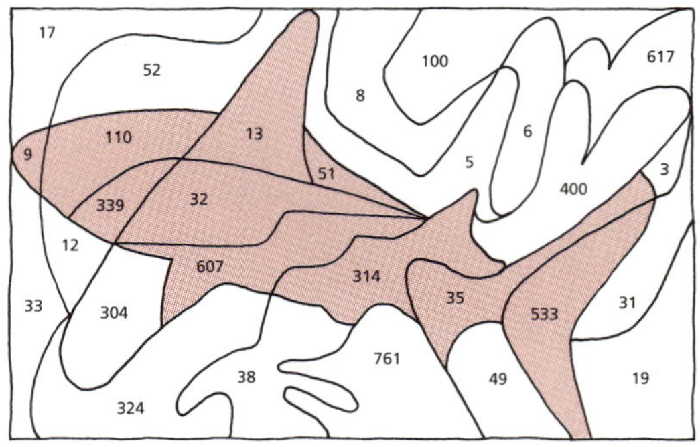

18 Elementary Math with *Pizzazz!*

(A SHARK)

▶ Add and subtract with regrouping

Name ________________

What Is a Carnivore's Favorite Game?

Find each sum or difference. Use the answers to complete the cross-number puzzle.

Across

1. 199 + 20 = **219**
3. 425 − 75 = **350**
5. 325 + 294 = **619**
7. 654 + 151 = **805**
9. 767 − 484 = **283**
10. 839 − 271 = **568**
11. 823 − 281 = **542**

Down

2. 333 − 152 = **181**
4. 820 − 270 = **550**
5. 193 + 482 = **675**
6. 456 + 472 = **928**
7. 273 + 562 = **835**
8. 934 − 372 = **562**

1. 2	2. 1	D 9		3. 3	4. 5	E 0
	H 8				L 5	
5. 6	W 1	6. 9		7. 8	0	8. 5
R 7	■	9. 2	8	A 3	■	6
10. 5	O 6	8		11. 5	S 4	T 2

Locate each number that appears in the code in the cross-number puzzle. You will use only the numbers that have a letter in their box. Write that letter above the number in the code.

S	W	A	L	L	O	W		T	H	E
4	1	3	5	5	6	1		2	8	0

L	E	A	D	E	R
5	0	3	9	0	7

Elementary Math with *Pizzazz!* 19

▶ Add and subtract with regrouping

Name ________________

Why Was the Math Teacher Sad?

Find each sum or difference. Then write the letter of the exercise above the matching answer at the bottom of the page.

E 23 + 8 = **31**	L 70 − 2 = **68**	U 119 + 21 = **140**	C 345 − 16 = **329**
B 100 − 90 = **10**	H 345 + 72 = **417**	R 304 − 130 = **174**	A 119 + 190 = **309**
P 222 + 88 = **310**	S 321 − 43 = **278**	Y 456 + 44 = **500**	N 444 + 157 = **601**
O 712 − 323 = **389**	M 323 − 246 = **77**	D 543 + 288 = **831**	T 148 − 49 = **99**

B	E	C	A	U	S	E		S	H	E
10	31	329	309	140	278	31		278	417	31

H	A	D		S	O		M	A	N	Y
417	309	831		278	389		77	309	601	500

P	R	O	B	L	E	M	S
310	174	389	10	68	31	77	278

20 Elementary Math with *Pizzazz!*

► Practice multiplication and division facts   Name ____________

Lost in a Cave

Help Travis find his way through the maze. Follow the path of answers for each multiplication or division fact.

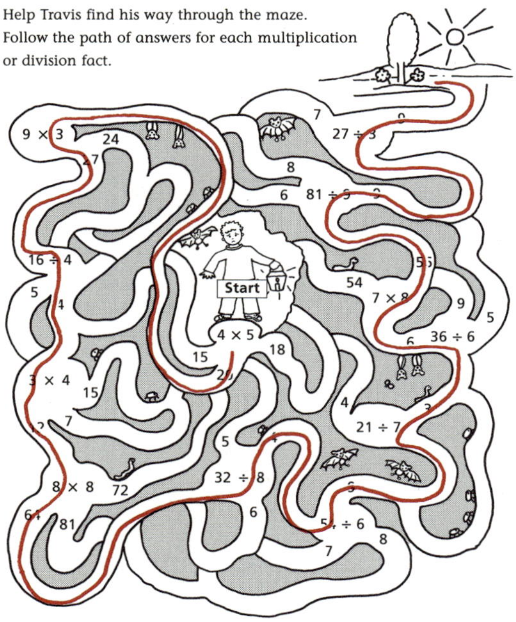

► Perform one-step division with remainders   Name ____________

Digging for Answers

Find each quotient and remainder. Use a ruler and draw lines to match answers that are the same. Your lines will cross out letters. Write the letters that remain in order at the bottom of the page.

37 ÷ 6 = 6 R1	46 ÷ 5 = 9 R1
14 ÷ 4 = 3 R2	13 ÷ 3 = 4 R1
28 ÷ 5 = 5 R3	55 ÷ 9 = 6 R1
7 ÷ 3 = 2 R1	5 ÷ 2 = 2 R1
9 ÷ 2 = 4 R1	70 ÷ 9 = 7 R7
47 ÷ 7 = 6 R5	23 ÷ 7 = 3 R2
82 ÷ 9 = 9 R1	33 ÷ 6 = 5 R3
63 ÷ 8 = 7 R7	53 ÷ 8 = 6 R5

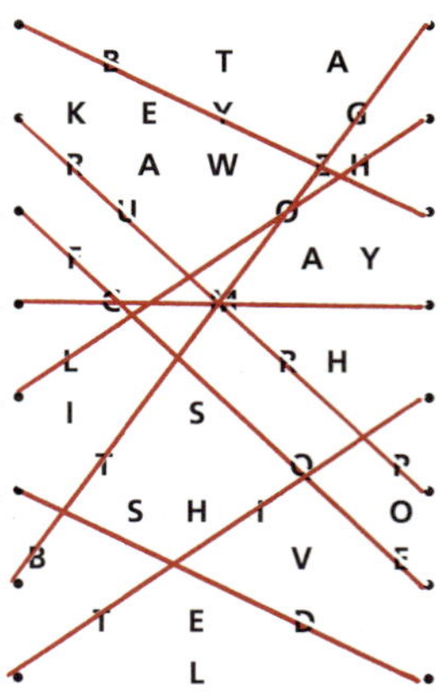

How do you make a dog stop digging?

TAKE AWAY HIS SHOVEL.

(TAKE AWAY HIS SHOVEL.)

► Multiply 2 digits by 1 digit without regrouping   Name ____________

Soaring with Multiplication

Find each product. Then connect your answers, in order, to complete the dot-to-dot.

1. 23 × 3	69
2. 11 × 6	66
3. 20 × 1	20
4. 45 × 1	45
5. 32 × 2	64
6. 13 × 2	26
7. 31 × 3	93
8. 28 × 1	28
9. 43 × 2	86
10. 10 × 4	40
11. 20 × 4	80
12. 11 × 5	55
13. 32 × 3	96
14. 25 × 1	25
15. 11 × 7	77
16. 30 × 3	90
17. 68 × 1	68
18. 42 × 2	84

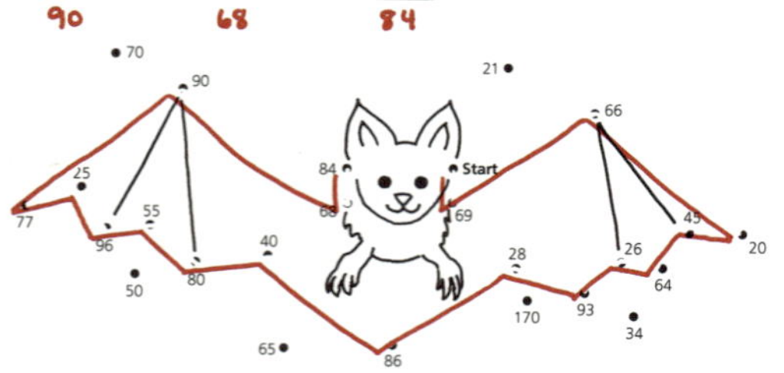

► Multiply 2 digits by 1 digit with regrouping   Name ____________

What a Day!

Find each product. Look for your answers in the story at the bottom of the page. Write the word of the problem in the space above the answer.

Julius	25 × 3	75
books	32 × 7	224
morning	18 × 4	72
forgot	16 × 5	80
when	39 × 2	78
home	13 × 4	52
sneezed	37 × 2	74
afternoon	43 × 9	387
finally	22 × 5	110
bus	49 × 2	98
day	56 × 2	112
Cara	68 × 3	204
homework	38 × 7	266
bell	28 × 5	140
knee	39 × 8	312
bad	45 × 2	90
playground	27 × 5	135
lunch	16 × 4	64
things	19 × 5	95
tooth	53 × 5	265

This MORNING (72) when I left for school, I FORGOT (80) my LUNCH (64). Then THINGS (95) really got BAD (90). I lost my HOMEWORK (266). I scraped my KNEE (312) on the PLAYGROUND (135). JULIUS (75) SNEEZED (74) on me. And WHEN (78) it was FINALLY (110) time to go HOME (52), I missed the BUS (98). I had a really bad DAY (112)!

► Divide 2 digits by 1 digit with remainders   Name

Mystery Picture Puzzle

Solve the division problems in the boxes at the bottom of the page. Then carefully cut out the picture pieces. Paste each piece on the box with the matching answer.

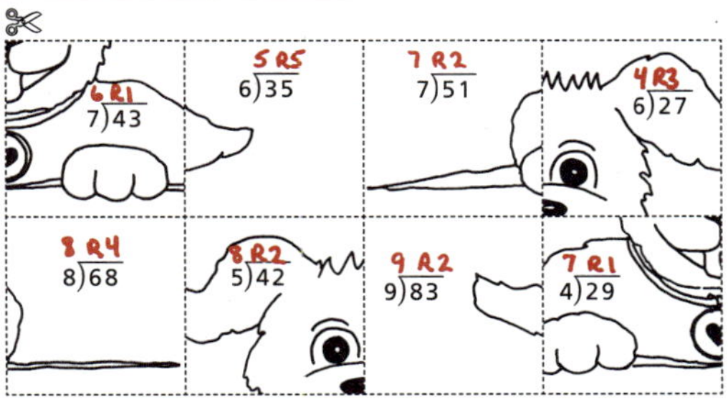

► Divide 2 digits by 1 digit with remainders   Name

Why Don't Lobsters and Crabs Share?

Divide. Find your answer in one of the boxes at the bottom of the page and cross out the box. Write the words that remain in order.

1. 8)62 **7 R6**
2. 9)89 **9 R8**
3. 7)38 **5 R3**
4. 9)38 **4 R2**
5. 4)21 **5 R1**
6. 5)42 **8 R2**
7. 8)33 **4 R1**
8. 6)57 **9 R3**
9. 9)75 **8 R3**
10. 4)26 **6 R2**
11. 9)47 **5 R2**

~~BECAUSE 7 R6~~	~~THEY 5 R3~~	BOTH 9 R1	~~ATE 8 R3~~
~~ALL 9 R8~~	~~OF 6 R2~~	ARE 7 R4	~~THEM 5 R1~~
~~WANTED 4 R2~~	JUST 6 R5	PLAIN 3 R2	~~TO 9 R3~~
~~BE 8 R2~~	SHELLFISH 7 R5	~~ALONE 4 R1~~	~~CLAWS 5 R2~~

BOTH ARE JUST PLAIN SHELLFISH

(BOTH ARE JUST PLAIN SELFISH)

► Use mental computation   Name

A Riddle for You

Find each answer.
Try to solve each exercise in your head.

- **M** The number of days in 2 weeks **14**
- **E** Double 5 and add 7. **17**
- **S** Start with 8. Add 3 and subtract 4. **7**
- **L** The number of tires on 2 cars **8**
- **O** $2 \times 2 \times 2 \times 2 =$ **16**
- **I** Start with 4. Add 6 and subtract 5. **5**
- **N** The number of fingers on 3 hands **15**
- **E** Triple 3 and add 3. **12**
- **T** $8 - 2 - 2 - 2 - 2 =$ **0**
- **G** Start with 20. Divide by 2 and add 1. **11**
- **O** The number of ears on 10 dogs **20**
- **A** Double 8 and add it to 8. **24**
- **N** $36 \div 6 \div 6 =$ **1**

What did the beach say when the tide came in?
Find each answer in the code. Write the letter of the exercise above it.

L	O	N	G		T	I	M	E		N	O		S	E	A
8	20	15	11		0	5	14	17		1	16		7	12	24

(LONG TIME NO SEE)

► Use mental computation   Name

Why Did the Cookie Go to the Doctor?

Part A Ring the letter of the operation you should use to solve the problem.

1. There are 8 boxes of crayons with 8 crayons in each box. How many crayons in all?
   (**C**) Multiplication **T** Subtraction **E** Division
2. Rob had 19 balloons. Three balloons popped. How many were left?
   A Addition (**Y**) Subtraction **C** Multiplication
3. Hannah had 35¢. She found a dime. How much money did she have then?
   F Subtraction **W** Division (**I**) Addition
4. There are 6 dogs and 48 dog biscuits. If shared equally, how many biscuits will each dog get?
   N Addition **O** Multiplication (**R**) Division
5. There are 5 shelves with 10 books on each shelf. How many books in all?
   P Division (**F**) Multiplication **S** Subtraction

Part B Ring the letter of the correct answer.

6. The number of inches in 2 feet	(**L**) 24	**E** 2	**R** 36
7. Triple 10 and then subtract 5	**D** 30	**P** 20	(**M**) 25
8. How many horns do 3 bulls have?	**A** 9	(**E**) 6	**I** 4
9. The number of thumbs on 5 hands	**Z** 8	**I** 6	(**U**) 5
10. Double 50 and subtract 25	**F** 125	(**T**) 75	**X** 50

Match each letter with a ring to the number of the exercise.

I	T		F	E	L	T		C	R	U	M	M	Y
3	10		5	8	6	10		1	4	9	7	7	2

► Choose reasonable answers Name

Nursery Rhyme Riddle

Estimate the cost of each purchase. Ring the letter that shows whether the child has **not enough** or **more than enough** money.

Paint box	$2.49	Eraser	$.90
Crayons	$1.99	Markers	$3.99
Glue	$.99	Scissors	$3.19
Notebook	$2.29	Paintbrushes	3 for $2.79
Folder	$.20	Pencils	2 for $.75
Ruler	$.59		

1. Judy needs to buy 2 pencils and a bottle of glue. She has $2.00.
   P not enough (C) more than enough
2. Mike has $7.00. He needs 1 package of markers, scissors, and 2 folders.
   (T) not enough D more than enough
3. Alita wants 1 notebook, a ruler, and a box of crayons. She has $4.00.
   (U) not enough B more than enough
4. Wayne needs a paint box, 3 paintbrushes, and a notebook. He has $8.00.
   A not enough (O) more than enough
5. Darius would like to buy 2 folders, 2 pencils, and a ruler. He has $1.50.
   (R) not enough S more than enough
6. Modupé has $4.00 to buy an eraser, 4 pencils, and a bottle of glue.
   E not enough (I) more than enough
7. Omar has $5.00. He needs a notebook, 3 folders, 2 pencils, and an eraser.
   N not enough (M) more than enough

What did Old Mother Hubbard say when she opened her cupboard?
Write the letter with a ring above the number of the exercise.

O I C U R M T
4 6 1 3 5 7 2

(Oh, I see you are empty.)

► Evaluate estimates Name

Check the Estimate

Decide if each estimated answer is an overestimate or an underestimate. Color overestimates red. Color underestimates blue.

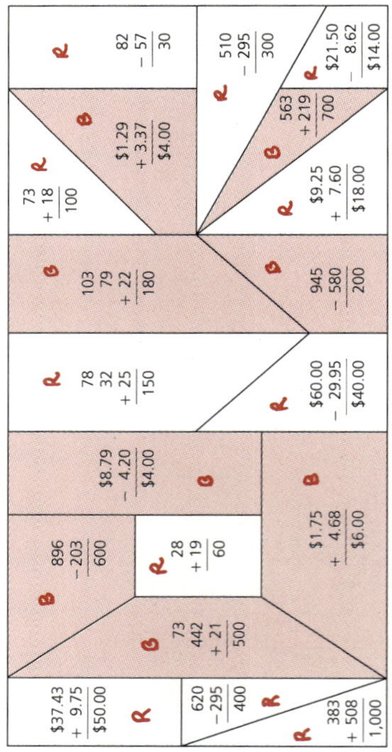

► Identify unit fractions Name

Those A-Maze-ing Fractions

Follow the path of fractions that name the part of each figure that is **not** shaded.

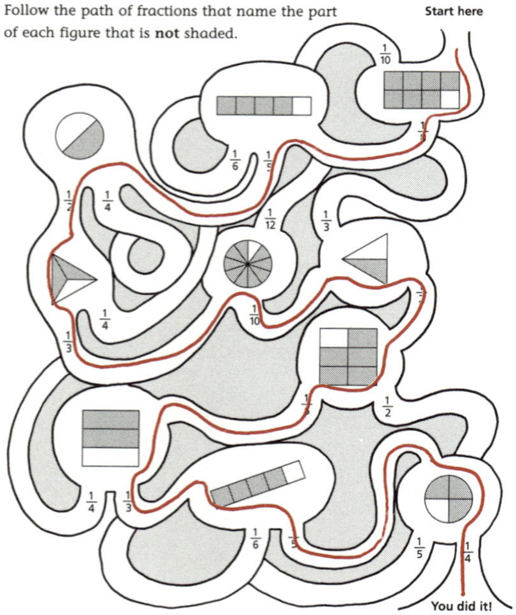

► Identify fractions Name

What Can You Break with Your Voice?

Follow the directions in each box. Then match the letter of the exercise with its answer in the code.

C Shade $\frac{3}{8}$

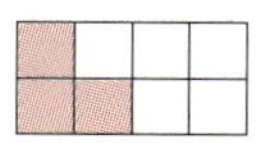

Write the fraction that is not shaded. $\frac{5}{8}$

I Shade $\frac{2}{5}$

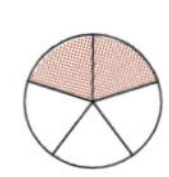

Write the fraction that is not shaded. $\frac{3}{5}$

L Shade $\frac{5}{6}$

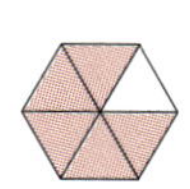

Write the fraction that is not shaded. $\frac{1}{6}$

S Shade $\frac{2}{3}$

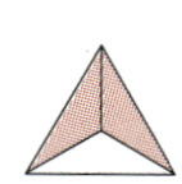

Write the fraction that is not shaded. $\frac{1}{3}$

E Shade $\frac{3}{10}$

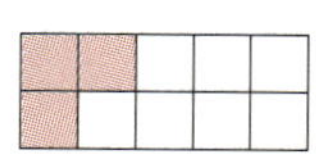

Write the fraction that is not shaded. $\frac{7}{10}$

N Shade $\frac{5}{8}$

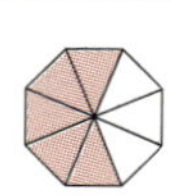

Write the fraction that is not shaded. $\frac{3}{8}$

S I L E N C E
$\frac{1}{3}$ $\frac{3}{5}$ $\frac{1}{6}$ $\frac{7}{10}$ $\frac{3}{8}$ $\frac{5}{8}$ $\frac{7}{10}$

► Compare fractions

Name ____________

What Rooms Have No Windows?

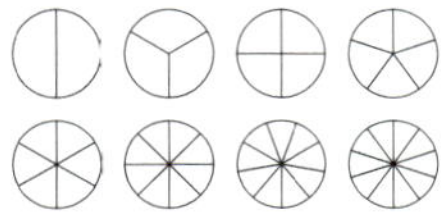

Compare the fractions. Write >, <, or = in the space between each pair.

1. $\frac{1}{3}$ > $\frac{1}{4}$ (S, R)
2. $\frac{2}{8}$ < $\frac{2}{5}$ (O, M)
3. $\frac{1}{8}$ < $\frac{3}{8}$ (T, R)
4. $\frac{3}{6}$ < $\frac{3}{4}$ (E, H)
5. $\frac{5}{9}$ > $\frac{2}{9}$ (U, A)
6. $\frac{6}{10}$ > $\frac{3}{10}$ (M, N)
7. $\frac{1}{2}$ < $\frac{2}{2}$ (G, S)
8. $\frac{5}{8}$ < $\frac{5}{6}$ (W, O)

Match the letter below the greater amount in each pair with the number of the problem.

M	U	S	H	R	O	O	M	S
2	5	7	4	3	8	8	6	1

► Find a fraction of a group

Name ____________

The Best Medicine

Solve. Then match the letter of the exercise with its answer in the code.

Mrs. Izzary baked 2 dozen (24) muffins. She frosted $\frac{1}{2}$ of them. She sprinkled $\frac{1}{4}$ of the muffins with powdered sugar, and she left $\frac{1}{4}$ of the muffins plain.

- T 12 muffins were frosted
- A 6 muffins had powdered sugar
- A 6 muffins were plain

Mrs. Izzary gave $\frac{1}{6}$ of the muffins she made to her neighbor.

- B 4 muffins were given to the neighbor
- R 20 muffins remained

The Izzarys ate $\frac{1}{4}$ of the remaining muffins for dessert that night.

- W 5 muffins were eaten for dessert
- L 15 muffins remained

The next morning, Mrs. Izzary packed $\frac{1}{3}$ of the remaining muffins in lunches.

- W 5 muffins were packed in lunches
- E 10 muffins remained

What should you drink when you are sick?

W	E	L	L		W	A	T	E	R
5	10	15	15		5	6	12	10	20

► Identify mixed numerals

Name ____________

Take a Bite Out of Mixed Numerals

Note: Use with page 36.

Write the mixed numeral that describes the part of each group that is shaded.

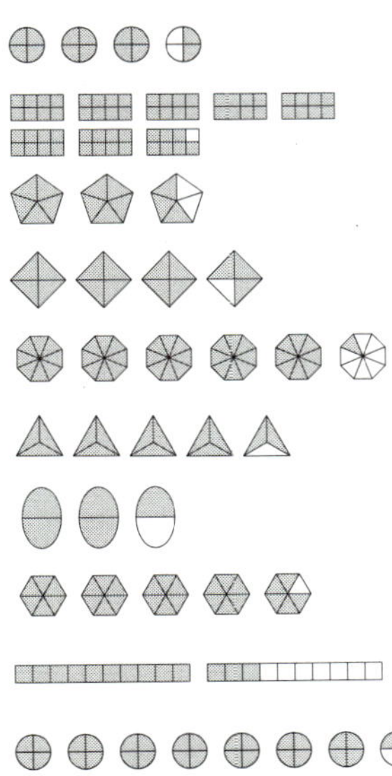

1. $3\frac{1}{2}$ →
2. $7\frac{7}{8}$ →
3. $2\frac{3}{5}$ →
4. $3\frac{3}{4}$ →
5. $5\frac{1}{8}$ →
6. $4\frac{2}{3}$ →
7. $2\frac{1}{2}$ →
8. $4\frac{5}{6}$ →
9. $1\frac{3}{10}$ →
10. $7\frac{1}{4}$ →

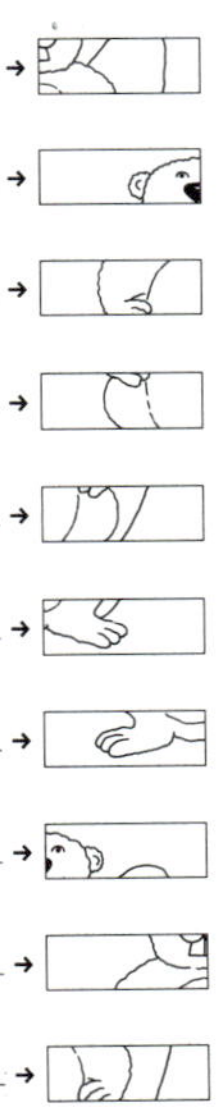

► Identify mixed numerals

Name ____________

Take a Bite Out of Mixed Numerals

Note: Use with page 35.

Carefully cut out the picture pieces. Match each picture piece to one shown on page 35 and note the mixed numeral that goes with it. Then paste the piece in the box below that has the same mixed numeral.

► Understand time

Name ____________

When Is the Best Time to See a Dentist?

Ring the letter of the correct answer. Use the calendar to answer questions 1–5.

January						
Sun.	Mon.	Tues.	Wed.	Thurs.	Fri.	Sat.
		1	2	3	4	5
6	7	8	9	10	11	12
13	14	15	16	17	18	19
20	21	22	23	24	25	26
27	28	29	30	31		

1. The 12th of January is
   S Monday E Sunday (T) Saturday
2. The date for the 3rd Monday of the month is the
   I 14th (O) 21st R 3rd
3. The first day of February is
   (O) Friday G Sunday M Tuesday
4. The date for the 1st Sunday of the month is the
   H 1st (T) 6th A 5th
5. The month of January ends on a
   T Saturday E Friday (H) Thursday
6. The last month of the calendar year is
   (H) December D September N July
7. The month right before June is
   F July B January (U) May
8. Hans started reading a book in March and finished 3 months later. He finished reading the book in
   (R) June T July M May
9. The 8th month of the year is
   Y July S September (T) August
10. The 4th month of the year is
    L May (Y) April N March

Write the letters of your answers in order.

TOOTHHURTY

Elementary Math with *Pizzazz!* 37

(TWO THIRTY)

► Calculate elapsed time

Name ____________

What Do You Do First Each Day?

Solve.

1. It took Ina 15 minutes to walk to school. If she left at 7:25 a.m., what time did she get to school? 7:40 a.m.
2. At 8:05 a.m., Ina's teacher said that a special visitor was supposed to arrive in 40 minutes. What time was the visitor expected? 8:45 a.m.
3. The visitor arrived 5 minutes late. He spent 30 minutes showing his collection of pet snakes. What time did he leave? 9:20 a.m.
4. The class spent the next 25 minutes writing thank-you notes to the visitor. What time did they finish their thank-you notes? 9:45 a.m.
5. Reading time started at 9:45 a.m. It lasted 1 hour and 15 minutes. When did reading time end? 11:00 a.m.
6. Art class started at 1:30 p.m. The class was 40 minutes long. What time was art class over? 2:10 p.m.
7. After school, Ina went to a friend's house. It took her 25 minutes to walk home from there. She got home at 5:00 p.m. What time did she leave her friend's house? 4:35 p.m.
8. Ina has 45 minutes worth of homework. It takes her 20 minutes to get ready for bed. If Ina's bedtime is 9 p.m., what is the latest time she can start her homework and still get to bed on time? 7:55 p.m.

Find each answer in the Code Box. Then match the letter in front of the answer to the number of the question in the code.

Code Box

Y	4:35 p.m.	A	9:20 a.m.	G	8:00 a.m.	W	7:55 p.m.	E	9:45 a.m.	R	2:30 p.m.
I	9:15 a.m.	U	7:40 a.m.	P	8:45 a.m.	S	8:45 p.m.	K	11:00 a.m.	O	2:10 p.m.

Y	O	U		W	A	K	E		U	P
7	6	1		8	3	5	4		1	2

38 Elementary Math with *Pizzazz!*

► Calculate elapsed time

Name ____________

It Is About Time

Solve.

W Chin took his dog for a walk. They left the house at 6:50 a.m. and returned at 7:40 a.m. How many minutes were they gone? 50 min

G Kelsey and her friend went to see a movie. The show started at 2:15 p.m. and ended at 3:55 p.m. How long was the movie? 1 hr, 40 min

N Becca's family started out for the museum at 1:05 p.m. The trip took 1 hour and 10 minutes. What time did they get to the museum? 2:15 p.m.

R Rosa baked cookies. The first batch went in the oven at 4:10 p.m. and came out at 4:22 p.m. How long did the cookies bake? 12 min

U Ethan went shopping with his mother. They left home at 10:30 a.m. and returned at 12:00 noon. How long were they gone? 1 hr, 30 min

K Randy's guitar lesson began at 4:15 p.m. If the lesson lasted for 30 minutes, at what time was the lesson over? 4:45 p.m.

M Sadie, the cat, finished eating 7:20 p.m. She had started eating 25 minutes before. What time was it when Sadie started eating? 6:55 p.m.

T Diane finished her science project at 3:00 p.m. She began working at 1:35 p.m. How much time did she spend working? 1 hr, 25 min

S A pizza was delivered to Bernard's house at 5:25 p.m. It had been ordered at 4:45 p.m. How long did it take for the pizza to arrive? 40 min

I Tracey was tired and decided to take a nap. She fell asleep at 1:45 p.m. and woke up at 2:55 p.m. How long did Tracey sleep? 1 hr, 10 min

What do you get when you cross a fish with an elephant?

Write the letter of the question above its answer.

S	W	I	M	M	I	N	G
40 min	50 min	1 hr 10 min	6:55 p.m.	6:55 p.m.	1 hr 10 min	2:15 p.m.	1 hr 40 min

T	R	U	N	K	S
1 hr 25 min	12 min	1 hr 30 min	2:15 p.m.	4:45 p.m.	40 min

Elementary Math with *Pizzazz!* 39

► Read temperature (°F)

Name ____________

Interesting Temperature Facts

Read each thermometer to find the temperature.

A You can tell the outside temperature by the number of times a cricket chirps. If it chirps 8 times in 15 seconds, the temperature is 75 °F.

C The hottest city in the United States is Yuma, Arizona. The average high temperature there is 90 °F.

O The temperature of a sea turtle's egg tells whether it will hatch as a boy or girl turtle. It will be a girl if the temperature is 85 °F.

E At the zoo in Phoenix, animals are given ice pops made of frozen fruits, vegetables, fish, and seeds to help keep the animals cool in Phoenix's usual summer temperature of 115 °F.

L Baby chicks must be kept very warm when they hatch. The best temperature to keep them healthy is 95 °F.

R Caves stay at about the same temperature all year. The temperature in most caves is around 50 °F.

Write the letter of the exercise above its answer.

R	E	A	L		C	O	O	L
50°	115°	75°	95°		90°	85°	85°	95°

40 Elementary Math with *Pizzazz!*

► Answers for pages 41–44

► Select appropriate customary units Name ____________

Measure Treasure

Travel through the maze by choosing the better unit to measure each thing described.

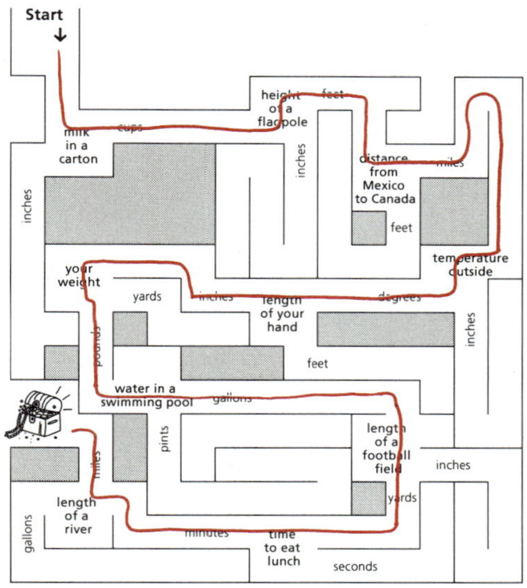

► Select appropriate metric units Name ____________

An Interesting Fact

Ring the letter of the better metric unit to use to measure each thing described.

1. Length of a playground	**(F)** meters	D liters
2. Amount of water in a glass	P millimeters	**(O)** milliliters
3. Weight of a cupcake	W centimeters	**(H)** grams
4. Length of a pencil	**(E)** centimeters	L kilometers
5. Weight of a watermelon	**(A)** kilograms	C grams
6. Amount of water in a fish tank	V centiliters	**(T)** liters
7. Distance from Japan to China	**(R)** kilometers	B centimeters
8. Height of a flagpole	G centimeters	**(S)** meters
9. Thickness of a dime	Y meters	**(J)** millimeters
10. Weight of a paper clip	**(M)** grams	U kilograms
11. Amount of medicine in an eyedropper	**(N)** milliliters	K milligrams

Match the letter of the answer to the number of the question in the code. When you are finished, you will find out the name of the U.S. President who was the first to want the United States to adopt the metric system.

T	H	O	M	A	S
6	3	2	10	5	8

J	E	F	F	E	R	S	O	N
9	4	1	1	4	7	8	2	11

► Estimate weight in metric units Name ____________

What Wears Shoes but Has No Feet?

Ring the letter of the better estimate.

1.  P 1 gram **(E)** 1 kilogram	2.  **(L)** 10 grams M 10 kilograms
3.  A 12 grams **(D)** 12 kilograms	4.  **(A)** 1 gram I 1 kilogram
5.  K 30 grams **(I)** 30 kilograms	6.  R 50 grams **(W)** 50 kilograms
7. 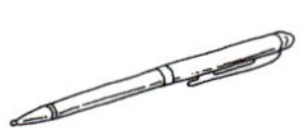 **(S)** 7 grams T 7 kilograms	8.  O 6 grams **(K)** 6 kilograms

Match the letter with a ring with the number of its box.

A		S	I	D	E	W	A	L	K
4		7	5	3	1	6	4	2	8

► Measure length in customary units Name ____________

What Has One Foot on Each Side and One in the Middle?

Measure each line to the nearer inch. Then write the letter of the line above your measurement in the code at the bottom of the page.

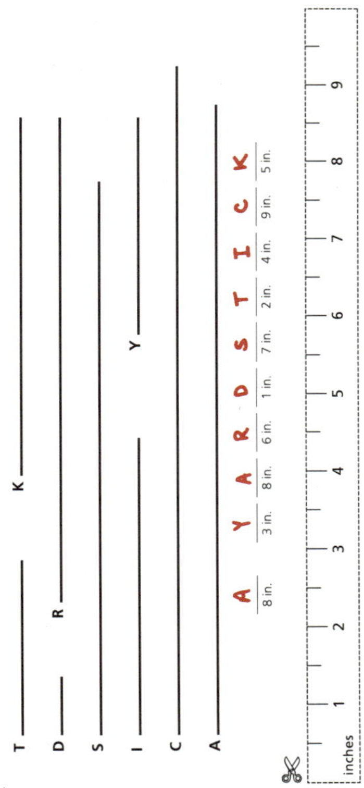

A		Y	A	R	D	S	T	I	C	K
8 in.		3 in.	8 in.	6 in.	1 in.	7 in.	2 in.	4 in.	9 in.	5 in.

► Measure length in metric units

Name

How Can You Recognize a Dogwood Tree?

Measure each line to the nearer centimeter. Then write the letter of the line above your measurement in the code at the bottom of the page.

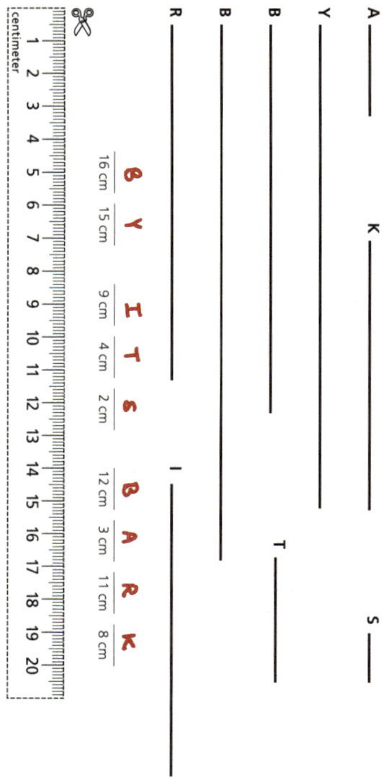

B	Y	I	T	S	B	A	R	K
16 cm	15 cm	9 cm	4 cm	2 cm	12 cm	3 cm	11 cm	8 cm

► Find perimeter

Name

Around You Go!

Find the perimeter of each figure. Then shade the part of the design that has your answer.

1. 12 in. by 2 in. rectangle — **28 in.**
2. 5 yd, 4 yd, 4 yd, 1 yd — **14 yd**
3. 6 m, 4 m, 4 m — **20 m**
4. 12 cm, 12 cm, 12 cm — **48 cm**
5. 10 ft, 2 ft, 10 ft — **24 ft**
6. 60 yd, 50 yd, 40 yd — **150 yd**
7. 3 in. each side — **24 in.**
8. 6 dm, 2 dm, 4 dm, 7 dm — **19 dm**
9. 4 in., 4 in., 4 in. — **12 in.**

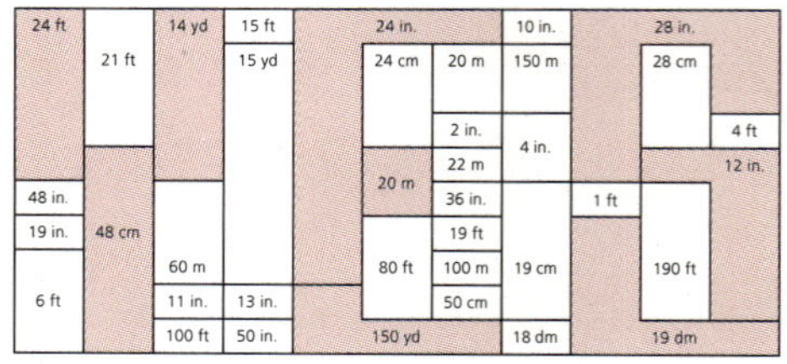

(YES)

► Find area

Name

Different but Equal

Find the area of each shaded part. Use a ruler and draw lines to match two areas that are the same. Your lines will cross out letters. Write the letters that remain in order at the bottom of the page.

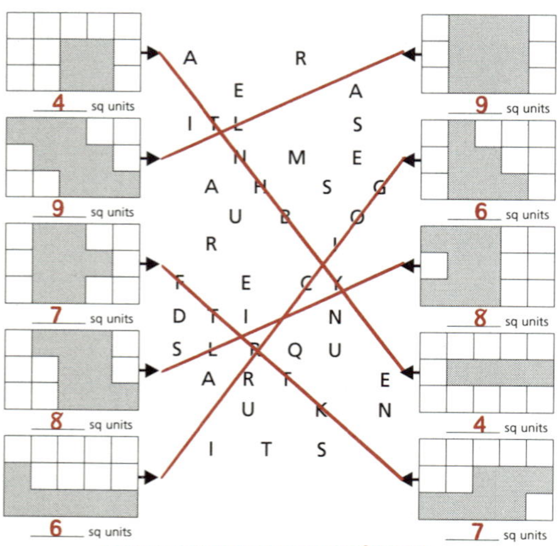

AREAISMEASUREDINSQUAREUNITS

(AREA IS MEASURED IN SQUARE UNITS)

► Find perimeter and area

Name

What Did the Guitar Say to the Rock Star?

Find the perimeter and area of each figure.

Example: □ perimeter = 4 units, area = 1 sq unit

1. perimeter = **10** units → I area = **4** sq units → K	2. perimeter = **20** units → M area = **9** sq units → G
3. perimeter = **14** units → O area = **12** sq units → E	4. perimeter = **16** units → C area = **7** sq units → T
5. perimeter = **22** units → P area = **11** sq units → U	6. perimeter = **18** units → Q area = **14** sq units → N

Match the letter with the measurement.

Q	U	I	T
18 units	11 sq units	10 units	7 sq units

P	I	C	K	I	N	G
22 units	10 units	16 units	4 sq units	10 units	14 sq units	9 sq units

O	N	M	E
14 units	14 sq units	20 units	12 sq units

▶ Determine volume Name

What Are They?

Write the number that tells how many cubes are used to make each figure. Use a ruler and draw lines to match two figures that have the same number of cubes. Your lines will cross out letters. Write the letters that remain in order at the bottom of the page.

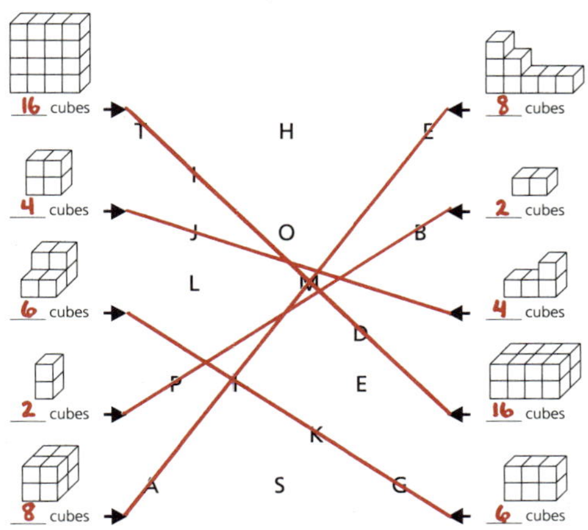

The more you take away from them, the bigger they get. What are they?

HOLES

▶ Identify faces of solid figures Name

What Do All 3-D Figures Have?

Which solid in each row can be made from the net shown? Ring the letter below that solid.

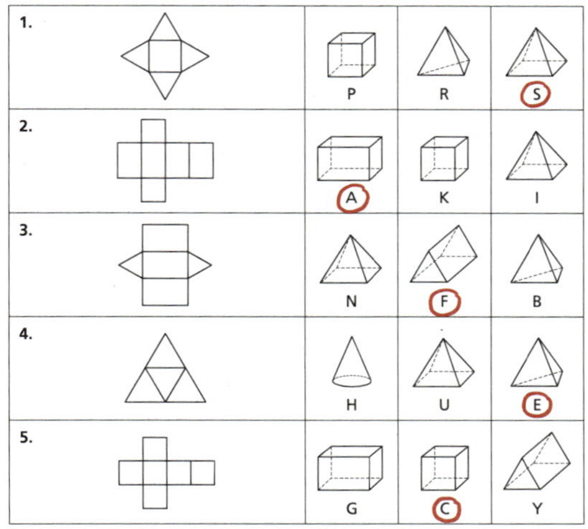

1.	P	R	(S)
2.	(A)	K	I
3.	N	(F)	B
4.	H	U	(E)
5.	G	(C)	Y

Write the letter with a ring above the number of its row.

F A C E S
3 2 5 4 1

▶ Identify attributes of 3-D shapes Name

Make It Right

Identify the figure that matches the description. Find the box at the bottom of the page that has the name of the figure in it. Ring the word above the name of the figure.

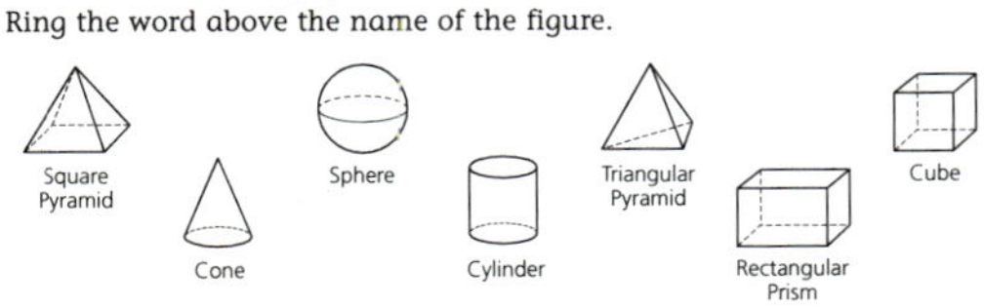

1. I have a circle on only one end. CONE
2. I have 6 faces. All of them are squares. CUBE
3. I do not have any flat surfaces. SPHERE
4. I have 5 faces. Four are triangular and one is square. SQUARE PYRAMID
5. I have two circle faces, one at each end. CYLINDER
6. I have 6 faces. All of them are rectangles, and two of them are squares. RECTANGULAR PRISM

Which is correct: "9 and 5 *is* 13" or "9 and 5 *are* 13"?

BOTH hexagonal prism	(NEITHER) rectangular prism	(NINE) sphere	PLUS triangular pyramid
(AND) square pyramid	IS oval	(FIVE) cylinder	THE rhombus
(ARE) cube	THIRTEEN octagonal prism	CORRECT trapezoid	(FOURTEEN) cone

Write the words with a ring in order.

NEITHER NINE AND FIVE ARE FOURTEEN

▶ Identify congruent shapes Name

Triangle Twins

Carefully cut out the nine triangles at the bottom of the page. Match each triangle cutout with a triangle at the top of the page that is exactly the same size and shape. Glue the cutout on top of the triangle it matches.

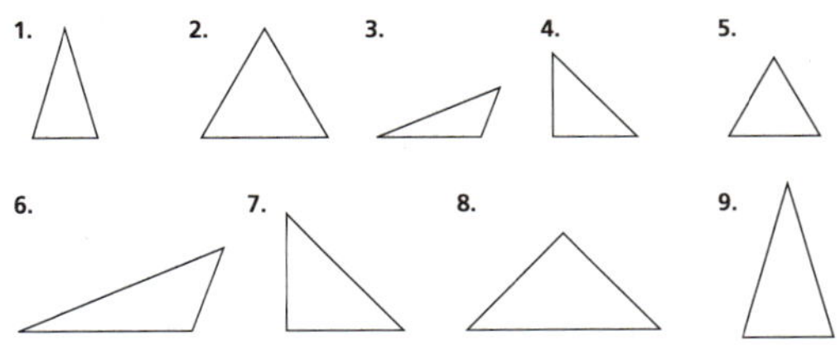

Write the letter found in each triangle in number order to spell out the word that tells how your pairs of triangles are related.

CONGRUENT

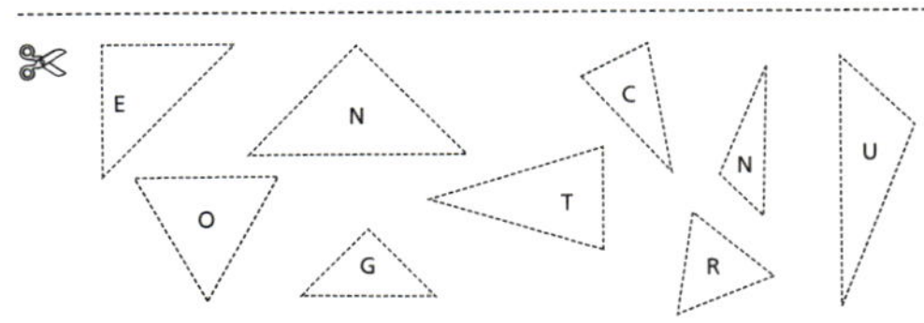

(NEITHER. NINE AND FIVE ARE FOURTEEN.)

► Recognize lines of symmetry Name ______

Is There Symmetry?

Decide if each figure has a line of symmetry.
Ring the letter in the column that tells your answer.

	Yes	No		Yes	No
1.	T	**(W)**	2.	**(A)**	H
3.	E	**(S)**	4.	**(I)**	H
5.	**(T)**	A	6.	**(A)**	D
7.	T	**(B)**	8.	**(A)**	W
9.	O	**(T)**	10.	**(I)**	M
11.	**(S)**	E	12.	H	**(A)**
13.	**(W)**	T			

Write the letters with a ring in order. The letters of your answer should read the same forward and backward.

WASITABATISAW

(WAS IT A BAT I SAW)

► Create flips Name ______

Reflect on Your Work

Shade squares in the box below each direction to show the flip.

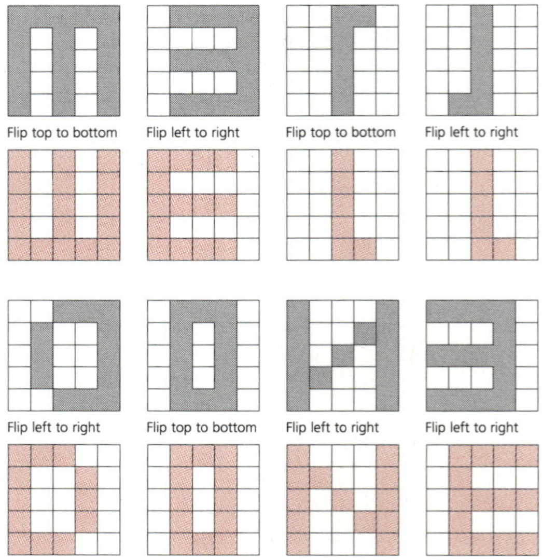

(WELL DONE)

► Identify points on a coordinate plane Name ______

How Do You Repair a Gorilla?

Write the letter of the point above its coordinates in the code.

W	I	T	H	A	M	O	N	K	E	Y
(3,1)	(4,4)	(5,2)	(3,4)	(1,5)	(3,3)	(5,3)	(4,3)	(1,3)	(5,5)	(5,1)

W	R	E	N	C	H
(3,1)	(3,2)	(5,5)	(4,3)	(3,5)	(3,4)

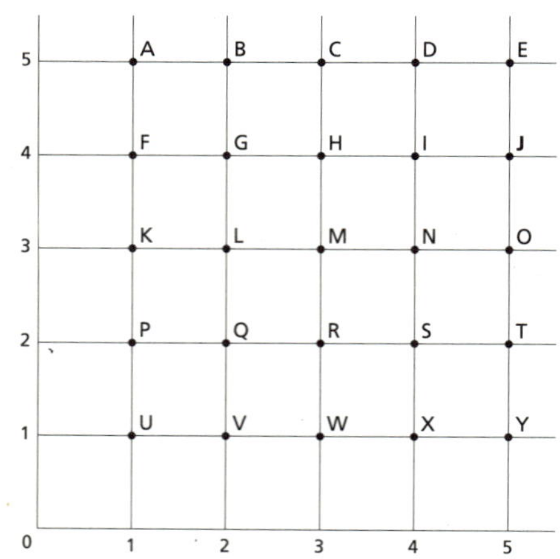

► Graph ordered pairs Name ______

Sail Through Graphing

First graph each point on the coordinate grid.
Be sure to label it with its letter. Then connect the points in A,B,C order.

A (6,5) **B** (14,8) **C** (7,17) **D** (6,4) Lift your pencil
E (5,16) **F** (1,12) **G** (1,8) **H** (6,6) **I** (4,11) **J** (6,17) Lift your pencil
K (1,5) **L** (4,4) **M** (14,4) **N** (13,2) **O** (3,2) **P** (2,4)

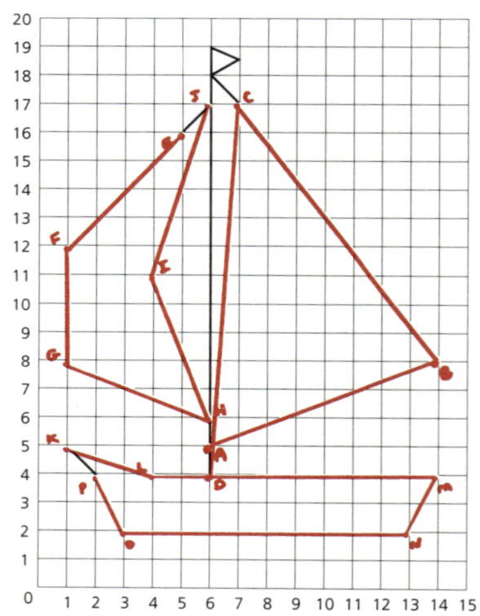

► Identify number patterns   Name

Something to Crow About

Identify the rule of each pattern.
Then follow the directions in the table and color.

Rule	Color
Add 3 or subtract 3	Orange
Add 9 or subtract 9	Red
Multiply by 5 or divided by 5	Brown
Add 7 or subtract 7	Blue
Multiply by 2 or divided by 2	Green

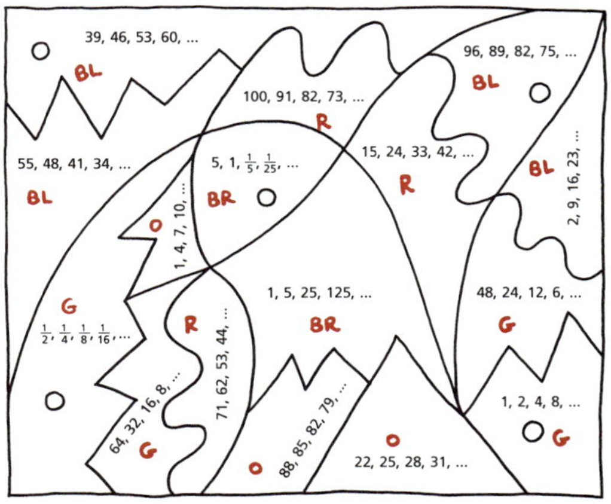

(ROOSTER)

► Extend number patterns   Name

What Is Black and White and Red All Over?

A separate pattern is shown with each path of arrows.
Follow the arrows and fill in the missing numbers.

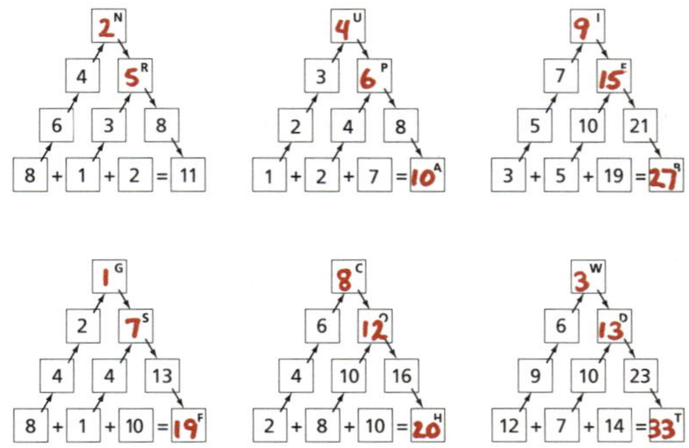

The numbers in the code are all numbers you have written in the pyramids. Each number has a letter in its box. Match the letter to the number in the code.

A		P	E	N	G	U	I	N		W	I	T	H		A		B	A	D
10		6	15	2	1	4	9	2		3	9	33	20		10		27	10	13

C	A	S	E		O	F		S	U	N	B	U	R	N
8	10	7	15		12	19		7	4	2	27	4	5	2

Permission is given to instructors to reproduce this page for classroom use. Copyright ©2003 Creative Publications, Wright Group/McGraw-Hill 0-3220-9690-1

► Extend number patterns   Name

What Always Ends Everything?

Follow the directions for each box. Write the numbers from the box in order on the lines below the box. Then fill in the final number. Be careful, the directions are not always the same.

1. Count forward.

   86, 116, 106, 126, 96

   86 96 106 116 126 136 — G

2. Count backward.

   1,998, 2,002, 1,996, 2,000, 1,994

   2,002 2,000 1,998 1,996 1,994 1,992 — L

3. Count forward.

   86, 66, 46, 126, 106

   46 66 86 106 126 146 — T

4. Count backward.

   1,287, 1,387, 1,337, 1,237, 1,187

   1,387 1,337 1,287 1,237 1,187 1,137 — E

5. Count forward.

   437, 837, 237, 637, 37

   37 237 437 637 837 1,037 — R

6. Count backward.

   5, 4, $5\frac{1}{2}$, $3\frac{1}{2}$, $4\frac{1}{2}$

   $5\frac{1}{2}$ 5 $4\frac{1}{2}$ 4 $3\frac{1}{2}$ 3 — H

Each answer has a letter below it.
Match the letter with the answer in the code.

T	H	E		L	E	T	T	E	R		G
146	3	1,137		1,992	1,137	146	146	1,137	1,037		136

Permission is given to instructors to reproduce this page for classroom use. Copyright ©2003 Creative Publications, Wright Group/McGraw-Hill 0-3220-9690-1

► Extend geometric growing patterns   Name

Growing with Patterns

Note: Use with page 61.

Draw the 4th figure in the pattern. Count how many squares you drew. Write that number in the table.

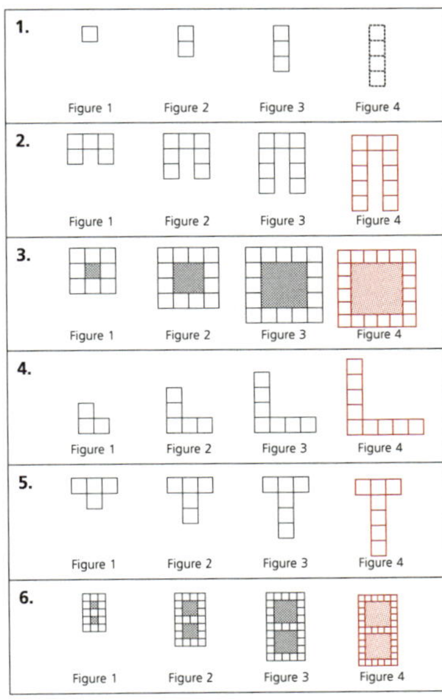

1.

Figure	Number of Squares
1	1
2	2
3	3
4	4 → F

2.

Figure	Number of Squares
1	5
2	7
3	9
4	11 → P

3.

Figure	Number of Squares
1	8
2	12
3	16
4	20 → S

4.

Figure	Number of Squares
1	3
2	5
3	7
4	9 → N

5.

Figure	Number of Squares
1	4
2	5
3	6
4	7 → Y

6.

Figure	Number of Squares
1	13
2	20
3	27
4	34 → R

Permission is given to instructors to reproduce this page for classroom use. Copyright ©2003 Creative Publications, Wright Group/McGraw-Hill 0-3220-9690-1

► Extend geometric growing patterns　　Name ______

Growing with Patterns

Note: Use with page 60.

7.

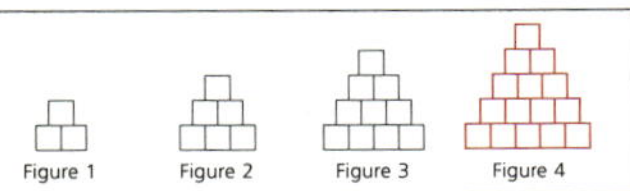

Figure	Number of Squares
1	3
2	6
3	10
4	15 → I

8.

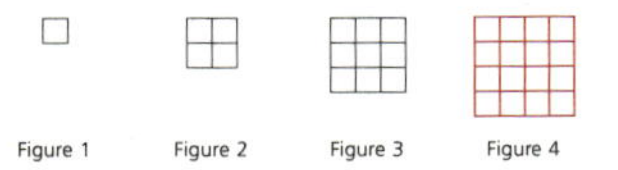

Figure	Number of Squares
1	1
2	4
3	9
4	16 → O

9.

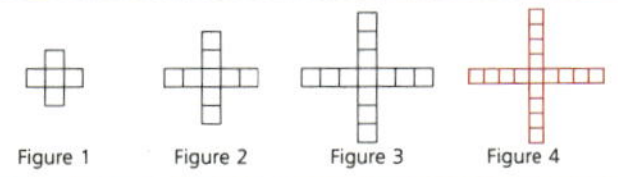

Figure	Number of Squares
1	5
2	9
3	13
4	17 → T

10.

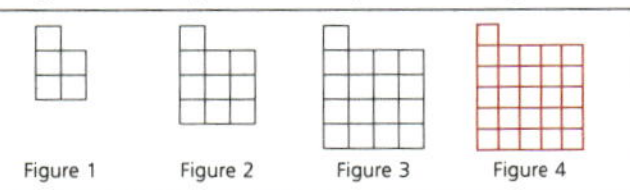

Figure	Number of Squares
1	5
2	10
3	17
4	26 → A

What has 100 legs but cannot walk?

Every number you wrote in a table has a letter after it. Match the letter with its number in the code.

F	I	F	T	Y		P	A	I	R	S
4	15	4	17	7		11	26	15	34	20

O	F		P	A	N	T	S
16	4		11	26	9	17	20

► Use symbols to represent quantities　　Name ______

The Missing Link

In each sentence a shape takes the place of a number. Find the value of that shape.

1. 3 + [9] = 12 □ = 9 (E)	**2.** (10) ÷ 5 = 2 ○ = 10 (E)
3. △4 + △4 + 6 = 14 △ = 4 (N)	**4.** [2] + [2] + [2] + 9 = 15 □ = 2 (L)
5. (20) ÷ 2 = 10 ○ = 20 (V)	**6.** △5 + 6 = 8 + 3 △ = 5 (O)
7. 9 × [6] = 54 □ = 6 (E)	**8.** (7) + (7) − 3 = 11 ○ = 7 (P)

What starts with "E" and ends with "E", but only has one letter?
Find each answer in the code. Write its letter above it.

E	N	V	E	L	O	P	E
6	4	20	9	2	5	7	10

► Recognize and solve number sentences　　Name ______

What Is Always in Front of You but Cannot Be Seen?

Ring the letter of the sentence that will correctly solve the problem. Then solve the problem. Find your answer in the code at the bottom of the page and write the letter with a ring above it.

1. Felipé ordered 36 peanut butter cookies and 24 chocolate chip cookies for the class party. How many cookies did he order in all?
   M 36 − 24 = □　(T) 36 + 24 = □ 60　Y 36 ÷ 2 = □
2. Kendra brought 4 packs of cups. Each pack had 12 cups. How many cups did she bring?
   R 12 − 4 = □　N 12 + 4 = □　(H) 4 × 12 = □ 48
3. Kiona needed to collect $30 for the party. So far, she has collected $26. How much more does she need to collect?
   L 30 + 26 = □　P 30 ÷ 26 = □　(R) 30 − 26 = □ 4
4. Eight children shared 2 pizzas. Each pizza had 12 slices. How many slices did each child get?
   (U) 24 ÷ 8 = □ 3　D 12 × 8 = □　M 12 ÷ 8 = □
5. The desks were arranged in groups of 4. There were 7 groups of desks. Someone sat at each desk. How many children were in the class?
   R 4 + 7 = □　(F) 4 × 7 = □ 28　A 7 − 4 = □
6. Telka made 18 party favors, and Caleb made 5 less than that. How many favors did Caleb make?
   (E) 18 − 5 = □ 13　I 18 ÷ 5 = □　W 18 + 5 = □

T	H	E		F	U	T	U	R	E
60	48	13		28	3	60	3	4	13

► Solve number sentences　　Name ______

Missing Number Dot-to-Dot

Find the missing number in each sentence.

- **A** 12 + [20] = 32
- **B** [24] ÷ 4 = 6
- **C** 11 × [6] = 66
- **D** 23 − [10] = 13
- **E** 9 + [22] = 31
- **F** 42 ÷ [7] = 6
- **G** [8] × 9 = 72
- **H** 41 − 11 = [30]
- **I** [18] + 112 = 130
- **J** [300] − 100 = 200
- **K** 2 × 7 + [3] = 17
- **L** [21] − 1 + 8 = 28
- **M** 3 × [9] + 9 = 36
- **N** [90] ÷ 10 = 9
- **O** [100] − 40 = 60
- **P** [41] + 59 = 100

Write the letter of the sentence above its answer in the Code Box.

Code Box

J	N	E	B	M	I	O	L	D	F	H	P	A	K	G	C
300	90	22	24	9	18	100	21	10	7	30	41	20	3	8	6

Complete the dot-to-dot. Follow the order of the letters in the Code Box.

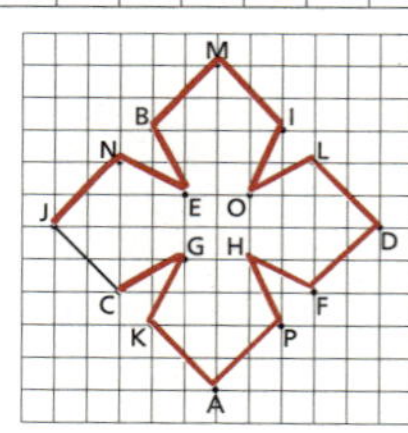

► Identify and extend patterns   Name ____________

What Could It Be?

Note: Use with page 66.

Write the missing numbers in each table.
Write the rule.

1. 1 → In | Rule: +3 | Out → 4

In	Out
3	6
5	8
7	10
9	**12**
11	14
13	**16**

2. 20 → In | Rule: **−5** | Out → 15

In	Out
100	95
90	85
80	75
70	**65**
60	55
50	45

3. 10 → In | Rule: **x 4** | Out → 40

In	Out
2	8
3	12
4	16
5	**20**
6	**24**
7	**28**

4. 13 → In | Rule: **+ 6** | Out → 19

In	Out
35	41
39	**45**
43	49
47	53
51	**57**
55	61

► Identify and extend patterns   Name ____________

What Could It Be?

Note: Use with page 65.

5. 24 → In | Rule: **÷ 2** | Out → 12

In	Out
12	6
22	11
32	16
42	**21**
52	26

6. 3 → In | Rule: **x 9** | Out → 27

In	Out
2	18
4	36
6	**54**
8	**72**
10	90

What is bright orange and sounds like a parrot?
Shade each section that has a number you wrote in one of the tables.

► Read graphs   Name ____________

What Can You Always Count On?

Answer the questions below each graph.
Then place the letter of each exercise above its answer in the code.

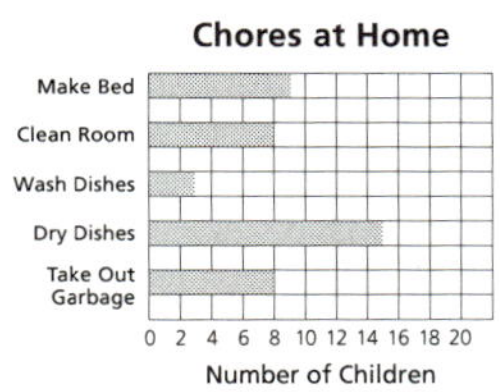

R How many children clean their room? **8**

F How many children either make their bed or clean their room? **17**

U If each child in this survey only named one chore, how many children participated in the survey? **43**

S How many more children make their beds than wash dishes? **6**

Y How many fewer children wash dishes than dry dishes? **12**

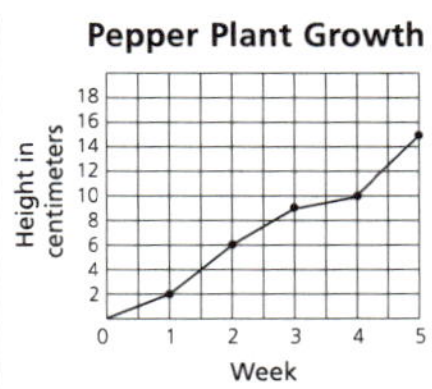

E How many centimeters did the pepper plant grow from week 1 to week 2? **4** cm

O What was the height of the pepper plant on week 4? **10** cm

I How many centimeters did the pepper plant grow from week 4 to week 5? **5** cm

G What was the least amount of growth during a one-week period for the pepper plant? **1** cm

N From week 1 to week 5, how many centimeters did the pepper plant grow? **13** cm

Y	O	U	R		F	I	N	G	E	R	S
12	10	43	8		17	5	13	1	4	8	6

► Make predictions   Name ____________

A "Most Likely" Maze

Suppose you either spin a spinner or pick a marble from a bag without looking. Follow the path of the result that would be most likely.

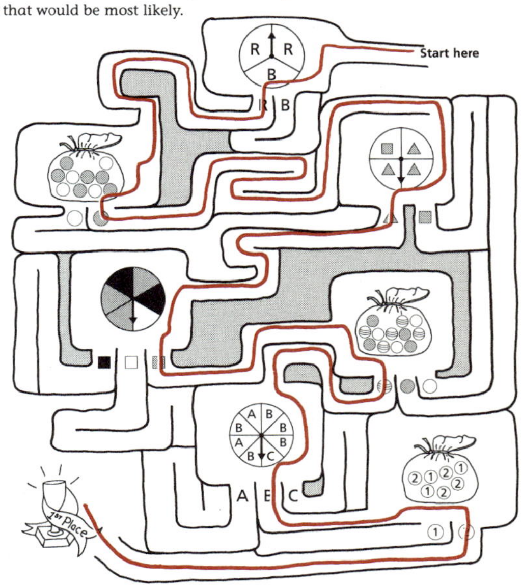

► Draw conclusions Name

What Do You Think?

Ring the letter in front of the correct answer.

1. △ ○ □ △ ○ □ The shape that comes next is (K) △ M ○	2. △ ÷ 2 = 12 (A) △ is greater than 10. R △ is less than 10.
3. 51, 53, 55, 57… L The next number will be even. (T) The next number will be odd.	4. 5 + □ > 10 O □ can be 5. (E) □ can be greater than 100.
5. The next figure will have (D) more than 20 dots Y fewer than 20 dots	6. △ × 9 = 108 G △ < 10 (I) △ > 10
7. odd number + even number = △ K △ is an even number. (U) △ is an odd number.	8. even number + □ = even number (P) □ is an even number. N □ is an odd number.
9. The bottom number is S odd (C) even	10. The shape on the missing piece is (Q) △ D ●

Did you hear about the duck that flew upside down?
Match the letter with a ring to the number of its exercise in the code.

I	T	Q	U	A	C	K	E	D	U	P
6	3	10	7	2	9	1	4	5	7	8

Elementary Math with *Pizzazz!* 69

► Determine likelihood of events Name

What Has Many Ears but Cannot Hear?

Use a ruler. Draw a straight line from each event to **more likely**, **less likely**, or **equally likely** to show the chances that the event will happen. Your lines will cross out letters. Write the letters that remain in order at the bottom of the page.

1. Omar rolls a number cube. He thinks the number that lands on top will be greater than 2.
2. Zena flips a coin. She thinks it will land heads up.
3. Nadia has a bag with 20 red cubes and 10 blue cubes. She thinks if she picks a cube without looking it will be red.
4. The school bus is late about once a week. Quan thinks the bus will be late on Monday.
5. There are 1 blueberry and 2 carrot muffins in a bag. Erika takes a muffin from the bag without looking. She thinks it will be a carrot muffin.
6. The names of 12 boys and 12 girls are placed in a box. Doug picks a name from the box without looking. He thinks he will pick a girl's name.
7. Peter packs a pickle with his lunch every day. He will pack a pickle with his lunch tomorrow.
8. Duane has 10 cards numbered from 1 to 10. Jody picks a card without looking. She thinks she will pick a 7.

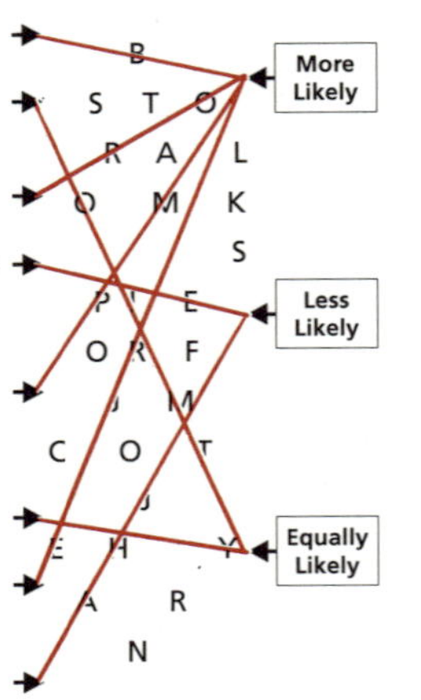

STALKSOFCORN

70 Elementary Math with *Pizzazz!*

(STALKS OF CORN)

► List possible outcomes Name

Think of What Is Possible

Follow the directions in each box. Find all possible answers.

1. 2, 3, 4 Multiply 2 of the numbers. Then add the third number to the product. 10, 11, 14	2. 1, 5, 9 Make 3-digit numbers. Use all 3 digits in each number you make. 159, 195, 519, 591, 915, 951
3. 2, 4, 6, 3, 0 Add only 2 of the numbers. Find the largest sum. 10 Find the smallest sum. 2	4. 4, 6, 8 Make the 3 largest numbers possible. Use all 3 in each number you make. 864, 846, 684
5. 2, 3, 7 Multiply 2 of the numbers. Then add the third number to the product. 13, 23, 17	6. 2, 4, 6 Add 2 of the numbers. Then multiply the sum by the third number. 36, 32, 20
7. 1, 2, 3, 4 Make the 3 smallest numbers possible. Use all 4 digits in each number you make. 1,234 1,243 1,324	

How can you tell if an elephant is getting ready to charge?
Possible answers for each exercise are grouped together. Ring only those that are correct. Then write the letters above the numbers with a ring in order.

S (H) R (E) (T) — 34 36 24 32 20
H (A) (K) (E) S — 3,124 1,234 1,324 1,243 1,342
(S) (O) D (U) — 10 11 12 14
(T) (H) (I) — 864 684 846
(S) (C) (R) (E) (D) (I) — 159 519 915 951 591 195
(T) G (C) (A) — 13 20 23 17
L (R) (D) — 9 10 2

HETAKESOUTHISCREDITCARD

Elementary Math with *Pizzazz!* 71

(HE TAKES OUT HIS CREDIT CARD)

► Determine probability of simple events Name

What Are the Chances?

Decide the probability of each event that is described. Write the probability as a fraction.

N You roll the cube. It will land with 5 on top. $\frac{1}{6}$	H You spin the spinner. It will land on Y $\frac{1}{4}$
R You pick a marble without looking. It will be a black marble. $\frac{3}{5}$	E You flip a coin. It will land heads up. $\frac{1}{2}$
D You spin the spinner. It will land on △. $\frac{1}{3}$	G You pick a tile without looking. It will **not** be a white tile. $\frac{2}{3}$
C You roll the cube. It will land with a number that is **not** 6 on top. $\frac{5}{6}$	S You pick a crayon without looking. It will be green. $\frac{1}{5}$
O You take a piece of fruit without looking. It will be an apple. $\frac{3}{4}$	A You roll the cube. It will land with a number that is less than 7 on top. $\frac{6}{6}$

Match the letter of the exercise with its answer in the code.

C	H	A	N	C	E	S	A	R	E	G	O	O	D
$\frac{5}{6}$	$\frac{1}{4}$	$\frac{6}{6}$	$\frac{1}{6}$	$\frac{5}{6}$	$\frac{1}{2}$	$\frac{1}{5}$	$\frac{6}{6}$	$\frac{3}{5}$	$\frac{1}{2}$	$\frac{2}{3}$	$\frac{3}{4}$	$\frac{3}{4}$	$\frac{1}{3}$

72 Elementary Math with *Pizzazz!*